The Energy Transition
and
The Struggle for Global Governance

The Energy Transition
and
The Struggle for Global Governance

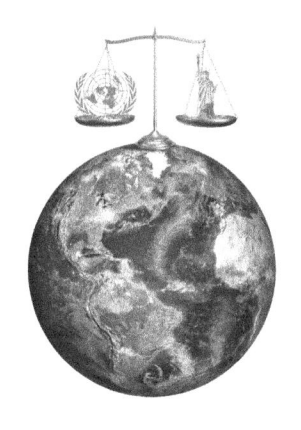

John R Fanchi

Texas Christian University, USA

World Scientific

NEW JERSEY • LONDON • SINGAPORE • BEIJING • SHANGHAI • HONG KONG • TAIPEI • CHENNAI • TOKYO

Published by

World Scientific Publishing Co. Pte. Ltd.
5 Toh Tuck Link, Singapore 596224
USA office: 27 Warren Street, Suite 401-402, Hackensack, NJ 07601
UK office: 57 Shelton Street, Covent Garden, London WC2H 9HE

Library of Congress Cataloging-in-Publication Data
Names: Fanchi, John R., author.
Title: The energy transition and the struggle for global governance /
　John R. Fanchi, Texas Christian University, USA.
Description: New Jersey : World Scientific, [2025] | Includes bibliographical references and index.
Identifiers: LCCN 2024017603 | ISBN 9789811290190 (hardcover) |
　ISBN 9789811290206 (ebook for institution) | ISBN 9789811290213 (ebook for individuals)
Subjects: LCSH: Energy industries--Environmental aspects. | Power resources--
　Environmental aspects. | Energy development--Environmental aspects. | Globalization.
Classification: LCC HD9502 .F36 2025 | DDC 333.79028/6--dc23/eng/20240527
LC record available at https://lccn.loc.gov/2024017603

British Library Cataloguing-in-Publication Data
A catalogue record for this book is available from the British Library.

Copyright © 2025 by World Scientific Publishing Co. Pte. Ltd.

All rights reserved. This book, or parts thereof, may not be reproduced in any form or by any means, electronic or mechanical, including photocopying, recording or any information storage and retrieval system now known or to be invented, without written permission from the publisher.

For photocopying of material in this volume, please pay a copying fee through the Copyright Clearance Center, Inc., 222 Rosewood Drive, Danvers, MA 01923, USA. In this case permission to photocopy is not required from the publisher.

For any available supplementary material, please visit
https://www.worldscientific.com/worldscibooks/10.1142/13759#t=suppl

Desk Editors: Balasubramanian Shanmugam/Steven Patt

Typeset by Stallion Press
Email: enquiries@stallionpress.com

Preface

The Energy Transition and the Struggle for Global Governance explains how the desire to achieve an energy transition is being used to catalyze a change in the world order as civilizations clash and evolve. A few years ago, an energy transition policy called the Goldilocks Policy was introduced to help guide the transition from fossil fuels to sustainable energy. The Goldilocks Policy was based on a historically reasonable timeframe to facilitate a peaceful energy transition by the end of the 21st century. Events around the world have shown that the energy transition is more likely to be chaotic rather than peaceful. This book presents evidence that the energy transition is being used to impose global governance over a new world order. It provides information that can help readers make informed decisions.

Part 1 presents a review of the historical emergence of the modern world order and discusses several factors that facilitated changes to the world order. Mechanisms for reshaping the existing world order during the energy transition are discussed in Parts 2–4:

- In Part 2, we explain how the environment can be weaponized to help establish a global government.
- In Part 3, we describe how the weaponization of climate change can be used to empower a global government.
- In Part 4, we show how global governance is being achieved by weaponizing institutions and events.

With this background, we describe a modern model of changing world orders. A selection of possible scenarios for a world order that might emerge after the energy transition is then presented.

John R. Fanchi
March 2024

About the Author

John R. Fanchi has a Ph.D. in physics and experience in academia and the energy industry. He has worked in several energy companies and has taught courses in energy and engineering at the Colorado School of Mines and Texas Christian University. He is the author of several books, including *Energy in the 21st Century, 5th Edition* (World Scientific, 2024), *Confronting the Enigma of Time* (World Scientific, 2023), *Reason, Faith and Purpose: The Ultimate Gamble* (World Scientific, 2022), *The Goldilocks Policy: The Basis for a Grand Energy Bargain* (World Scientific, 2019), *Principles of Applied Reservoir Simulation, 4th Edition* (Elsevier, 2018), *Introduction to Petroleum Engineering* (with R.L. Christiansen, Wiley, 2017), *Integrated Reservoir Asset Management* (Elsevier, 2010), *Math Refresher for Scientists and Engineers, 3rd Edition* (Wiley, 2006), *Energy: Technology and Directions for the Future* (Elsevier-Academic Press, 2004), *Shared Earth Modeling* (Elsevier, 2002), *Integrated Flow Modeling* (Elsevier, 2000), and *Parametrized Relativistic Quantum Theory* (Kluwer, 1993). He was co-founder of the International Association for Relativistic Dynamics and served as its President from 1998–2004. He is a Distinguished Member of the Society of Petroleum Engineers and co-edited the General Engineering volume of the *Petroleum Engineering Handbook* published by the Society of Petroleum Engineers.

Contents

Preface   v

About the Author   vii

Part 1   How Did We Get Here?   1

Chapter 1   Emergence of the Modern World Order   3

Chapter 2   What Paths Are Possible?   25

Part 2   Weaponizing Environmentalism   39

Chapter 3   Environmental Socialism   41

Chapter 4   Fabian Socialism   51

Chapter 5   Road to Globalism   57

Chapter 6   U Thant Weaponizes the Environment   65

Part 3   Weaponizing Climate Change   69

Chapter 7   What Is Anthropogenic Climate Change?   71

Chapter 8   Is the Climate Change Debate Settled?   83

Chapter 9   Maurice Strong and Global Environmental Socialism   91

Chapter 10   Obama Picks Up the Environmental Baton   103

Part 4   Achieving Global Governance **111**

Chapter 11   Weaponizing the Privileged Minority   113

Chapter 12   Weaponizing National Governments   131

Chapter 13   Weaponizing Human Conflicts   147

Chapter 14   Weaponizing a Pandemic   173

Chapter 15   What Does the Future Look Like?   205

Appendix: The Goldilocks Policy for Energy Transition   215

References   221

Index   241

Part 1
How Did We Get Here?

Chapter 1

Emergence of the Modern World Order

The Goldilocks Policy for energy transition was introduced a few years ago to provide a general framework for transitioning from fossil fuels to sustainable energy in a reasonable time frame (Fanchi and Fanchi, 2015; Fanchi, 2019). The rate of transition was set based on the length of time needed to achieve transitions from wood to coal and from coal to oil. Historical data from the United States indicated that each energy transition lasted approximately 60 years. The rate of transition was estimated to be 2% per year and the energy transition from fossil fuels to nuclear fission and renewable energy would be completed before the end of the 21st century. An update of the Goldilocks Policy is presented in the Appendix.

The Goldilocks Policy called for a historically reasonable timeframe to be used to allow society to responsibly move from an infrastructure based primarily on fossil fuels to an infrastructure based on sustainable energy. Events around the world, however, have shown that the energy transition is more likely to be chaotic rather than peaceful.

Our goal in this book is to show that the energy transition is serving to catalyze a change in the world order as civilizations clash and evolve. We explain how the modern world order emerged in Chapter 1 of Part 1. We then describe different factors that can catalyze changes to the existing world order in Chapter 2.

1.1 Historical Origin of the Modern World Order

Those who cannot remember the past are condemned to repeat it

—Santayana (1905–1906)

A key element of the modern world order is the concept of sovereign state. This concept emerged from the resolution of the Thirty Years' War (1618–1648). It is a good place to begin our review of the history leading to the modern world order (Haass, 2020).

1.1.1 Thirty Years War and the Treaty of Westphalia

The Thirty Years War took place primarily in central Europe and affected much of Europe (BOE Thirty Years War, 2023). It concluded in 1648 with the Treaty of Westphalia. The treaty presaged the development of the international system because it established the principle of state sovereignty, a fundamental principle of international law. The principle of state sovereignty recognizes that sovereign states should accept the independence and boundaries of other sovereign states.

The Treaty of Westphalia played a significant role in creating the current international system consisting of sovereign states (BOE Westphalia, 2023). As a rule, a sovereign state exercises authority over activities within its territory. Sovereign states, also known as countries or nations, should obey the following rules:

1. Govern themselves without external interference.
2. Do not interfere in the internal affairs of other sovereign states.
3. Respect the territorial borders of other sovereign states.
4. Do not attempt to forcibly change territorial borders.

Acceptance and implementation of these rules brought more peace and stability between sovereign states but did not eliminate war.

A period of relative peace and stability in Europe was ended by the emergence of Napoleon Bonaparte in France, whose expansionist ambitions sparked wars across Europe. Austria, Prussia, Russia, and the United Kingdom formed an alliance that defeated Napoleon. The victors met in Vienna with the defeated French, excluding Napoleon, in 1814 and 1815 to negotiate a settlement. The resulting Congress of Vienna established the Concert of Europe which incorporated France into the new international order (BOE Congress of Vienna, 2022).

The Concert of Europe preserved existing dynasties and opposed revolutionary movements. However, it was unable to survive the unification of Germany by Prussian Minister-President Otto von Bismarck in 1871.

1.1.2 *Colonialism and the Meiji Restoration*

European countries, Japan, and the United States colonized many areas around the world, including the Middle East, South Asia, East Asia, and the Americas. The colonization was motivated by economic considerations, as well as national pride and the quest for glory.

In the 19th century, China was subject to divisive imperial rule, domestic resistance to central authority, and foreign aggression from other countries. The Opium Wars were an attempt by China, under the rule of the Qing Dynasty, to prohibit opium trade with European merchants because of the damaging effects of opium use (BOE Opium Wars, 2023). China fought the British in the first Opium War (1839–1842), and the British and French in the second Opium War (1856–1860). The foreign powers demanded concessions from China. As a result, the West opened trade with China and subjected China to Western influence. A series of incursions by the British, French, Germans, Japanese, and Russians followed the Opium Wars.

The first Opium War (1839–1842) marked the beginning of a period known as the century of humiliation to the Chinese. The century of humiliation did not end until Mao Zedong proclaimed the founding of the People's Republic of China in 1949. During this period, China reconsidered its role in the world, especially relations between sovereign states. Today, China believes that internal dissension and weakness invite foreign aggression. The Chinese Communist Party uses the century of humiliation to justify a central government that can preserve national unity and sovereignty.

Japan did not interact with other nations at the beginning of the 19th century. The United States, another nation with a Pacific coastline, was looking for new markets when American Commodore Matthew Perry led four ships into Tokyo Bay in 1853. Japan had expelled most foreigners in 1639 and Perry was seeking to re-establish regular trade and discourse with Japan. In addition to the advanced firepower on his ships, Perry brought gifts intended to show Western cultural superiority, including a telescope, telegraph, working model of a steam engine, and wines and liquors. He left a letter seeking a trade agreement with the Japanese government.

Perry was unable to secure an agreement with the Japanese in his 1853 visit. In 1854, he returned to Japan with a larger squadron of ships. Confronted with superior firepower, the Japanese reluctantly agreed to sign the Treaty of Kanagawa on March 31, 1854. The treaty provided limited maritime cooperation and access to ports but did not open trade. It did create opportunities for future contact between Japan and the United States, and eventually trade between the two nations.

Japanese rulers were primarily military leaders, or shoguns, in the latter half of the 12th century. Emperors appointed shoguns who then served as de facto rulers. Western efforts to open Japan resulted in a successful political challenge to the ruling shogunate, or military government. Emperor Meiji restored imperial rule in 1868. Young samurai from feudal areas that were traditionally hostile to the shogunate were among the leaders of the Meiji Restoration (BOE Meiji Restoration, 2023). The threat of foreign encroachment was one of the principal concerns of the revolutionaries. The Meiji Restoration marked the beginning of Japan's modernization and emergence as a major world power.

Meiji was emperor of Japan from 1867 to 1912. The modernization of Japan from a feudal country into a great power occurred during Meiji's reign. Japan established a bureaucratic government, implemented industrial policies, and built a modern military like those in Europe and the United States. In addition, Japan expanded its territorial control over parts of Korea, Taiwan, and China. The 1904–1905 Russo-Japanese War was fought between Japan and Russia over competing territorial claims in Manchuria and Korea. This was the first war in the modern era in which an Asian power, Japan, defeated a European power, Russia.

The Japanese defeat of Russia in 1905 ended the Russo-Japanese War (1904–1905) and Russia's desire to establish hegemony over all of Asia. It also contributed to domestic unrest. An uprising in 1905, now known as the Revolution of 1905, was instrumental in convincing Tsar Nicholas II to issue the October Manifesto, a document that would replace unlimited autocracy with a constitutional monarchy. The Tsar's policies and sporadic dissolution of the Duma, or Russian parliament, which was a consequence of the Revolution of 1905, spread dissatisfaction with Russian rule among ethnic minorities within the Russian Empire (BOE Russian Revolution, 2023).

1.1.3 *World War I and the Treaty of Versailles*

Many European settlements along the Atlantic coast of North America were British colonies during the 18th century. Dissatisfaction with British taxation motivated the British colonies to forcibly resist British rule in

1775 and declare their independence from Britain in 1776. This colonial struggle for independence is known as the Revolutionary War or the American War of Independence. A new country, the United States of America, emerged from the conflict.

The United States expanded from a coastal nation of thirteen states to a continental nation with 48 states stretching from the Atlantic Ocean to the Pacific Ocean by the early 20th century. The United States became a significant player on the global stage when its allies needed its economic and military power during World Wars I and II.

The June 1914 assassination of Archduke Franz Ferdinand of Austria in Sarajevo is typically considered the beginning of World War I. Recurring conflicts between Russia and Austria–Hungary in the Balkans, the mobilization of militaries, and the growth in power of neighboring Germany under the leadership of Prussian chancellor Otto von Bismarck at the end of the 19th century contributed to the outbreak of war. Successors to Bismarck were more willing to accept risk and be aggressive toward other sovereign states.

World War I pitted the Central Powers, notably Germany, Austria–Hungary, and the Ottoman Empire (now known as Turkey), against the Allies. The Allies included France, Great Britain, Russia, Italy, Japan, and the United States (BOE World War I, 2023). The United States entered World War I in April 1917 at a time when trench warfare in Europe had resulted in a stalemate between combatants.

Prior to the entry of the United States into the war, ships were transporting supplies from the United States to Britain and France to aid their war effort. Germany decided to engage in unrestricted submarine warfare to thwart the resupply of its enemies, but the decision also led to the torpedoing of the British-registered ocean liner RMS Lusitania by a German U-boat on May 7, 1915. The sinking of the Lusitania turned public opinion in the United States against Germany and indirectly supported United States President Woodrow Wilson's decision to enter the war against Germany.

Hostilities ended in November 1918 after Germany asked President Wilson to arrange an armistice in October 1918. A conference was convened in Paris to draft a peace agreement. Allied leaders representing the United Kingdom (Prime Minister David Lloyd George), France (Prime Minister Georges Clemenceau), the United States (President Woodrow Wilson), and Italy (Prime Minister Vittorio Orlando) dominated the conference. Other allies and defeated nations, including Germany, were not influential in drafting the peace treaty now known as the Treaty of Versailles (BOE Versailles, 2023). The Treaty took effect on June 28, 1919, five years after the assassination of Archduke Ferdinand. Germany reluctantly agreed to accept blame for the war as part of the treaty, make reparations and other concessions to allied nations, and limit its military capability.

The Treaty of Versailles included the Covenant of the League of Nations, the first supranational organization of the modern era. Earlier supranational organizations included the Catholic Church and the Holy Roman Empire, a precursor of the European Union.

Members of the League of Nations agreed to respect state sovereignty. The United States played a significant role in starting the League of Nations, but never became a member. The United States Senate rejected the proposed treaty in November 1919. Membership would have made the United States a founding member of the League of Nations. The absence of the United States as a member of the League of Nations weakened the League. Woodrow Wilson and the League of Nations are discussed in the context of globalism in Part 2.

1.1.4 Setting the Stage for World War II

Empires such as the Austro-Hungarian, Ottoman, and Russian empires ended because of World War I. The global map changed when an empire fell. For example, many modern Middle Eastern countries were formed when British and French officials redrew the map of the Middle East following the defeat of the Ottoman Empire. In addition, President Wilson's support of self-determination encouraged the spread of nationalism around the world. Nationalism can be considered a sentiment that arises when people in a certain region come "to see themselves as sharing a distinct identity, the result of a common history, language, religion, ethnicity, and/or set of political beliefs" (Haass, 2020, p. 19).

Germany

Germany created the Weimar Republic, a parliamentary democracy, after World War I. The Great Depression in the United States adversely affected the Weimar Republic. The Great Depression began in 1929 when reckless speculation and inadequate regulation resulted in a stock market crash in the United States. The crash led to a sharp decline in American productivity, a rise in unemployment, and bank failures.

The Great Depression exemplified the failure of capitalism and democracy and contributed to economic problems in Germany, such as inflation, and the collapse of the Weimar Republic. Consequently, a worried German populace chose to accept the suspension of civil liberties in exchange for the promise of stability and prosperity. National socialism replaced the parliamentary democracy in Germany.

In 1932, the National Socialist, or Nazi, Party had become the largest political party in the German parliament. Socialists wanted the means of producing and distributing goods to be owned collectively or by a centralized government that was authorized to control the economy. In 1933, Nazi Adolf Hitler became the chancellor of Germany and consolidated power in a centralized government. Hitler dismantled democratic protections and began to rearm the military despite constraints on the role of the German military specified by the Treaty of Versailles. The Nazi Party took on the characteristics of a fascist organization: it sought to centralize authority under a dictator who imposed stringent socioeconomic controls and glorified nationalism.

Russia

The Revolution of 1905 was the first of two revolutions in early 20th century Russia (BOE Russian Revolution, 2023). The Russian army was not well-prepared for its 1914 entry into World War I. It had some successes but also defeats. The public was unhappy with massive casualties, retreats, and economic hardships including food shortages. It was also dissatisfied with unsubstantiated rumors of official incompetence and treason.

The second revolution in 1917 began in February (of the Julian calendar; March in the Gregorian calendar) when factory workers went on strike for higher wages but were ignored by the ruling class. The demonstrators moved toward the center of the capital city in Petrograd (now known as Saint Petersburg). They were joined by more laborers and university students. The government tried to disperse the demonstrators by force. Several members of the Imperial Guard revolted and joined the demonstrators. Petrograd was taken by the revolutionaries. The tsarist monarchy was at risk.

The Petrograd Soviet of Workers' and Soldiers' Deputies was formed and took control of a provisional government. The first executive committee included supporters of Vladimir Lenin and believed that the Russian Revolution was the beginning of a global socialist revolution, but they did not have enough political strength to control the unstable provisional government. The provisional government had assumed responsibility for governing but did not have the power to rule. Furthermore, it was unable to balance its desire to play a defensive role with the demands of its World War I allies in a war that was still raging. The Russian empire began to separate into independent states such as Finland and Ukraine.

The Romanov Dynasty, led by Nicholas II and the Royal Family, came to an end. The House of Romanov was the reigning imperial house of Russia from 1613 to 1917. The Royal Family was arrested, incarcerated for several months, and executed in July 1917.

After months of instability, the Bolsheviks, a revolutionary faction of the Marxist Russian Social Democratic Labour Party, led by Vladimir Lenin, took control of the government and installed Lenin as Premier in October 1917 (in the Julian calendar; November 1917 in the Gregorian calendar). Lenin formed a government in a new Soviet state and built an army to defend it. The Russian Revolution of 1917 led to the replacement of tsarist rule with a Marxist government.

Joseph Stalin succeeded Lenin in 1924. Stalin instituted 5-year plans that attempted to rapidly industrialize the Soviet Union, collectivize agriculture, and suppress dissent. His policies resulted in widespread famine and economic hardship. He imprisoned and executed millions of people — estimates range from 10 million to over 20 million — who were considered threats to Stalin's regime.

Japan

The Great Depression in the United States during the 1930s affected nations around the world. Japan needed large imports of food to feed its growing population and natural resources such as oil. To pay for imports, Japan needed to export goods, but tariffs imposed by Western countries limited exports. Japan attempted to solve its economic problems by seeking military conquests.

China

The Qing Dynasty in China lasted from 1644 to 1912. The modern Chinese populace was not satisfied with the Qing Dynasty because the dynasty was unable to modernize the country nor defend its sovereignty. The fall of the Qing Dynasty marked the end of imperial rule in China. Chinese Nationalists, a non-Communist authoritarian movement, took control of key regions of the country. Chinese Nationalist rule was challenged in the 1930s when Japan invaded and occupied parts of China.

Italy

Benito Mussolini, the fascist leader of Italy, invaded Abyssinia (now Ethiopia) in 1935. Abyssinia appealed to the League of Nations for help. The League of Nations imposed economic sanctions that were ineffective in stopping the aggression. Italy conquered Abyssinia. Other nations noticed that Italy did not appear to face significant consequences for its aggression.

League of Nations and Appeasement

The lack of effective action by the League of Nations on behalf of China and Abyssinia showed the world that the League would tolerate aggressive behavior between sovereign states with little resistance. The practice became known as appeasement. The League of Nations later accepted German annexation of European regions such as Austria in 1938. These events were part of the lead-up to World War II.

The aggressive behavior of Germany, Japan, and Italy alarmed many people. Winston Churchill, for example, was a British politician who disagreed with those who supported acts of appeasement. Churchill formed a new British government in 1940 with the aim of resisting German advances.

1.1.5 *World War II and Nuclear Weapons*

Germany, Japan, and Italy formed an alliance in World War II known as the Axis powers. Their governments did not respect the sovereignty of other nations, which was a key element of the prevailing world order. Other nations around the world were not prepared to resist the Axis powers.

Military readiness required both military capability and the willingness to use it. The absence of suitable military readiness in nations coveted by the Axis powers, reliance on ineffective international agreements, and appeasement of German aggression in the 1930s paved the way for the outbreak of World War II.

World War II began in earnest when Adolf Hitler ordered German forces to invade Poland on September 1, 1939. Prior to its invasion of Poland, Germany signed a non-aggression pact with the Soviet Union in August 1939. The pact allowed Germany to proceed with the conquest of Western Europe without the risk of a two-front war. In exchange, the pact gave the Soviet Union the Baltic States, parts of Poland, and time to prepare for war with Germany. The pact was short-lived, however. Hitler violated the pact in June 1941 when he ordered the German invasion of the Soviet Union.

In the early years of the war, the United States remained officially neutral. However, President Franklin D. Roosevelt was sympathetic to the nations resisting German aggression and attempted to support them without directly entering the war. He implemented the Lend-Lease program to supply food, oil, and other materials to the Allies, including the United Kingdom, the Soviet Union, France, and China. At the same time, Roosevelt sought to maintain domestic political support, especially from people who opposed American involvement in the war. In addition, Roosevelt's government imposed a selective embargo against Japan to prevent Japan from acquiring the resources it needed for its military buildup.

The Japanese attack on the United States naval base at Pearl Harbor, Oahu, Hawaii on December 7, 1941, catalyzed the United States declaration of war against the Axis powers. The leading Allied nations were the United States, the United Kingdom, and the Soviet Union.

The Allied victory in Europe over Italy and Germany was completed in the spring of 1945. Led by the United States, the Allies liberated most of Asia and the Pacific from Japan by the summer of 1945. The Allies had to decide between invading the home islands of Japan or using newly developed nuclear weapons called atomic bombs. An international collaboration of scientists developed atomic bombs in the Manhattan Project. The atomic bomb is based on nuclear fission, or the release of energy associated with the splitting of a large nucleus into smaller nuclei and neutrons.

Harry Truman, who became president of the United States following Roosevelt's death in April 1945, had to decide if atomic bombs should be used against Japan. Truman authorized their use in the Japanese cities of Hiroshima (August 6) and Nagasaki (August 9) in August 1945.

Japan agreed to cease hostilities within days of the second bombing. Japanese officials signed surrender documents on the USS Missouri on September 2, 1945, ending the Pacific War and World War II.

The use of nuclear weapons ushered in the nuclear age. The Allies had decisively defeated the Axis powers. Leaders of Allied nations could have adopted the World War I model and chosen to punish the Axis powers, but they wisely chose to treat the defeated Axis powers with respect. For example, United States Secretary of State George C. Marshall proposed a European self-help program in 1947 that became known as the United States-sponsored European Recovery Plan. The European Recovery Plan was widely known as the Marshall Plan. It was used to help rebuild European economies, especially in countries outside of the sphere of influence of the Soviet Union, from 1948 to 1951 (BOE Marshall Plan, 2023). As a further sign of goodwill, Germany, Italy, and Japan were subsequently integrated into multinational institutions and agreements. Respect for sovereignty became a key component of the United Nations Charter in 1945, which stated in Principle 1 of Article 2 in Chapter 1 that "1. The Organization is based on the principle of the sovereign equality of all its Members" (UN Charter, 1945).

The United States emerged as a major world power during the first half of the 20th century based on its performance in the two World Wars. Indeed, nations around the world considered the United States a superpower after the war because of its military prowess, including the acquisition of nuclear weapons, and an intact economy.

1.2 The Nuclear Age

The Nuclear Age began during World War II. The development and use of nuclear weapons during wartime was the first application of nuclear reactions. Today, countries that have nuclear weapons now treat them as key components of their national arsenals. Some countries are willing to live without nuclear weapons, often because they cannot afford a nuclear arsenal. In some cases, they have alliances with countries that have nuclear weapons. Other countries without nuclear weapons covet them as a means of deterrence and self-defense.

The history of the Nuclear Age is reviewed here because of its importance to the modern world order. We begin with the history of wartime and peacetime applications and then highlight the electromagnetic pulse (EMP), a key component of a nuclear weapon.

1.2.1 History of the Nuclear Age

In 1895, German physicist Wilhelm Roentgen discovered a brand-new type of radiation known as X-rays. Roentgen's X-rays could penetrate a person's body and produce an image of their internal structure. Henri Becquerel, a Frenchman, made the discovery of radioactivity in 1896 while examining the fluorescence of a uranium salt for Roentgen's X-rays. The discovery of a new radioactive element was initially reported in 1898 by French physicist and physician Marie Curie and her husband Pierre Curie. Marie Curie was originally from Warsaw, Poland. The name "polonium" was chosen to honor Marie's homeland.

The "rays" that radioactive elements emit were named alpha, beta, and gamma by Ernest Rutherford. Today, we are aware that the beta ray is an electron, the gamma ray is an energetic photon, and the alpha ray is the helium nucleus. Rutherford and his colleagues discovered the atomic nucleus by bombarding tiny metallic foils with alpha particles in 1913 at the Cavendish Laboratory in Cambridge. The scattering of alpha particles by atoms within the foil revealed that most of the atomic mass is concentrated in its core. The concentration of mass is called the nucleus of the atom. Two components were found in the nucleus: a positively charged proton, and a new, electrically neutral particle known as the neutron. James Chadwick discovered the neutron in 1932 while working in Rutherford's laboratory.

In 1934, physicist Leo Szilard proposed the idea of a neutron chain reaction. Szilard was aware that radioactive materials could produce additional neutrons when the materials interacted with neutrons. A chain reaction could occur if the density of the radioactive material and the neutron count were high enough. When a process releases neutrons that trigger further reactions, those additional reactions in turn release more neutrons, resulting in a chain reaction.

Szilard envisioned two uses for the neutron chain reaction: a peaceful harnessing of the reaction to produce consumable energy, and the explosive release of energy for use in a hostile environment. Szilard filed patents such as the patent presented in Figure 1.1 after realizing the potential importance of chain reactions. He hoped his patents would stop, or at least hinder, the widespread development of military applications of chain reactions. Szilard's patents were the first attempt in human history to control the proliferation of nuclear technology.

PATENT SPECIFICATION 630,726

Application Date: June 28, 1934. No. 19157/34.
„   „   July 4, 1934. No. 19721/34.

One Complete Specification left (under Section 16 of the Patents and Designs Acts, 1907 to 1946): April 9, 1935.

Specification Accepted: March 30, 1936 (but withheld from publication under Section 30 of the Patent and Designs Acts 1907 to 1932)

Date of Publication: Sept. 28, 1949.

Index at acceptance: —Class 39(iv), P(1:2:3x).

PROVISIONAL SPECIFICATION
No. 19157 A.D. 1934.

Improvements in or relating to the Transmutation of Chemical Elements

Figure 1.1 British Chain Reaction Patent Filed by Leo Szilard in 1934

Low-energy ("slow") neutrons were used to bombard radioactive material for the first time in 1935 by Italian physicist Enrico Fermi and his colleagues in Rome. The Fermi unit, abbreviated as fm in particle physics, is the femtometer with a length equal to 10^{-15} m. It is named after Enrico Fermi. The range of the nuclear force that binds a nucleus together is one Fermi, which is about equal to the diameter of a nucleus. Fermi's experiments were explained in 1938 by Lise Meitner and Otto Frisch in Sweden and Otto Hahn and Fritz Strassmann in Berlin as a fission process.

The fission process occurs when a larger nucleus splits into smaller nuclei and releases energy The fission reaction $U^{235} + n \rightarrow Ba^{141} + Kr^{92} + 3n$ is shown in Figure 1.2. It illustrates the collision of a neutron n with a uranium nucleus U^{235} (uranium-235) to yield the fission products Ba^{141} (barium-141) and Kr^{92} (krypton-92) and three neutrons. If the three neutrons produced in the reaction are allowed to interact with other uranium nuclei, the first fission reaction provides enough neutrons to support three additional reactions. The fission process, if allowed to continue, can result in an explosive chain reaction as neutrons continue to interact with available uranium nuclei.

Fermi successfully operated the first sustained chain reaction at the University of Chicago on December 2, 1942, approximately one year after the United States entered World War II. His reactor, known as Chicago Pile-1 (CP-1), was the first manmade nuclear reactor. The demonstration

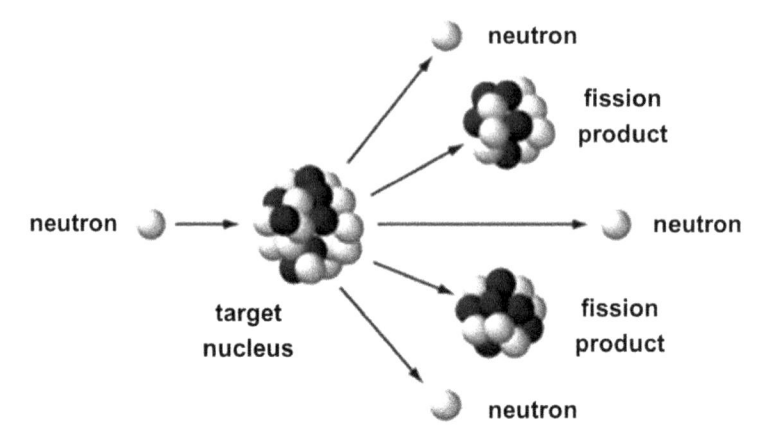

Figure 1.2 Illustration of the Fission Reaction $U^{235} + n \rightarrow Ba^{141} + Kr^{92} + 3n$

that a chain reaction was sustainable was a precursor to the development of nuclear fission weapons.

The Manhattan Project

Many well-known German scientists left their country and joined the Allied effort to create the first nuclear weapons in the United States. The Allied effort, known as the Manhattan Project, resulted in the successful development of a bomb based on nuclear fission. The resulting device was generally referred to as an atomic bomb, though the phrase "nuclear bomb" is more appropriate. In 1945, the first nuclear weapon was set off in a desolate area close to Alamogordo, New Mexico.

By the spring of 1945, Hitler's Germany was in ruins and there was no reason to employ the new weapon in Europe. However, Japan was still fighting the Allied forces in the Pacific and did not appear ready to submit until it had suffered a defeat at home. Many lives would have been lost in an amphibious attack against the Japanese islands, including Japanese combatants as well as non-combatants. Harry Truman, who had acceded to the presidency following President Roosevelt's death, decided that the new weapon could shorten the war and save lives. Fearing that a demonstration would not impress the wartime leaders of Japan, Truman ordered the use of the weapon on Japanese cities.

The Enola Gay, an American bomber named after Enola Gay Tibbets, mother of its pilot Colonel Paul Tibbets, dropped the first nuclear bomb ever used in combat on the Japanese city of Hiroshima on August 6, 1945. Approximately 90% of the city was destroyed, and almost 130,000 people died. The Japanese government refused to surrender, and a second nuclear bomb was detonated over the Japanese city of Nagasaki on August 9, 1945. The shock of two nuclear attacks and a Soviet invasion of Manchuria on August 9, 1945, impelled the Japanese Emperor to accept the terms of surrender from the Allies on August 14, 1945, and to formally surrender aboard the USS Missouri (Figure 1.3) on September 2, 1945.

News of the use of nuclear weapons spread throughout the world. The use of nuclear weapons led to mass destruction and did not discriminate between combatant and non-combatant. People began to realize that nuclear weapons had significantly changed the consequences of future total wars.

Figure 1.3 The USS Missouri Berthed at Pearl Harbor, Hawaii (© Fanchi, 2004)

The Nuclear Arms Race

Nuclear power has both peacetime and wartime applications. Peacetime applications of nuclear fission were developed in the 1950s. Wartime applications included the nuclear fission bombs detonated by the United States over Japan, and more powerful nuclear fusion bombs developed in the 1950s. Nuclear fusion bombs are also known as thermonuclear or hydrogen bombs.

A nuclear arms race began in the early 1950s when the Soviet Union acquired the technology to produce nuclear weapons. The United States held a slight lead in strategic nuclear capability over the Soviet Union in 1972 because the United States had a four-to-one advantage in deliverable warheads.

The quality of the missiles used as delivery vehicles by the United States nuclear arsenal increased significantly between 1972 and 1982. The United States benefited from rocket technology developed during the Space Race between the Soviet Union and the United States in the 1960s. The Space Race was driven by President John F. Kennedy's 1961 call to land a man on the Moon by the end of the decade. The first Moon landing occurred in July 1969.

The Soviet Union attained rough nuclear parity with the United States by 1982. Nuclear parity was intended to achieve deterrence, that is, the idea that neither side in a dispute would risk enduring unacceptably large losses in life and property by choosing to use nuclear weapons. Deterrence underlies the policy of Mutual Assured Destruction (MAD), which states that all participants in a nuclear conflict would risk the destruction of their own societies. MAD presumes that all societies involved in a dispute must be vulnerable to one another. If one society achieves a strategic advantage over the others, the more vulnerable societies may choose to strike first. A lack of parity increases the possibility of instability and war.

Ronald Reagan, President of the United States during the 1980s, proposed the Strategic Defense Initiative (SDI), commonly known as "star wars" after a trilogy of popular science fiction movies. The purpose of SDI was to defend the United States from attack from intercontinental ballistic missiles (ICBMs). It would require the development of a missile defense system that could intercept ICBMs at various stages of flight. The SDI missile defense system could also deter rogue nations with small nuclear arsenals from using nuclear weapons without regard to MAD. SDI was an alternative to MAD but put the balance of nuclear power at risk.

Emergence of the Modern World Order 19

SDI required substantial investment in a new missile system that required an economy that could support the increase in military spending. The economy of the Soviet Union was unable to compete with the development of a new missile system in the United States. The Union of Soviet Socialist Republics dissolved into separate states in the late 1980s. Russia and Ukraine are two examples of former Soviet republics that became independent states. They also possessed nuclear weapons. Ukraine agreed to give up its nuclear arsenal in 1994 when it signed the Budapest Memorandum. In return, the United States, Russia, and Britain agreed to "respect the independence and sovereignty and existing borders of Ukraine" and to "refrain from the threat or use of force" against Ukraine (Pifer, 2014). Russian incursion into Ukraine in 2014 to take Crimea and another in 2022 are clear examples that Russia is not upholding the terms of the Budapest Memorandum.

1.2.2 *The Electromagnetic Pulse*

The detonation of nuclear weapons emits an EMP, which can endanger electrical networks on a broad geographic scale (Fanchi, 2024). Figure 1.4 is a visualization of several prospective threats (US CBO Electric Grid, 2020). The horizontal axis represents an assessment of the potential economic damage of each threat, and the vertical axis represents an estimate of the frequency of occurrence. Table 1.1 summarizes the likelihood of occurrence, which ranges from very low to very high.

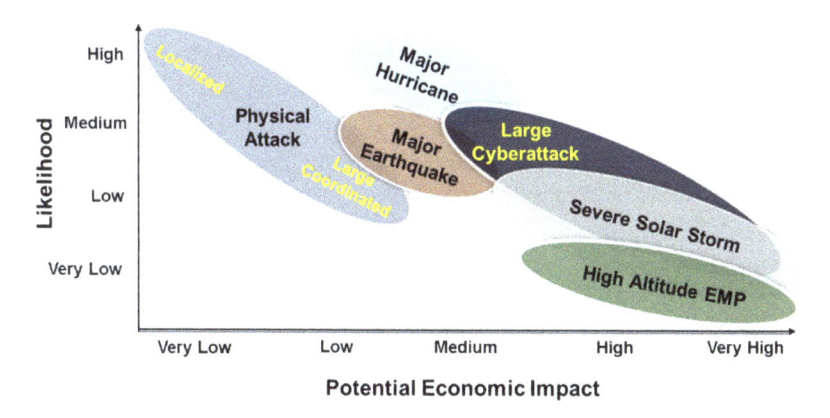

Figure 1.4 Potential Threats to Electrical Grids

Table 1.1 Estimating Potential Threats to the Grid (US CBO Electric Grid, 2020)

Measure	Potential Economic Impact	Likelihood (Expected Average Occurrence)
Very low	Hundreds of millions of $ or less	≈1 every 150 years or more
Low	Billions of $ or less	≈1 every 100 years
Medium	Tens of billions of $	≈1 every 50 years
High	Hundreds of billions of $	≈1 every 10 years
Very high	Trillion $ or more	

Most of the potential threats in Figure 1.4 are events that commonly occur during an average lifetime. Natural disasters include earthquakes, hurricanes, and solar storms. People are responsible for physical assaults, cyberattacks, and EMPs. The potential impact of an EMP emitted by the detonation of a nuclear bomb above the Earth merits closer examination.

Although the possibility of an EMP attack is considered remote, Russian President Vladimir Putin has threatened to deploy nuclear weapons in the 2022 conflict with Ukraine. The range of an EMP depends on the height of the explosion of a nuclear weapon. A surface explosion has a range that is less than the range of an airburst, or elevated explosion.

The impact of an EMP has three components, or periods (US CBO Electric Grid, 2020). Component 1 is a pulse that can harm electronics and computer systems. Component 2 is a lightning-like pulse. If the lightning defenses of the grid are still intact after interacting with Component 1, the lightning defenses can mitigate the effect of Component 2. Component 3 is the effect with the longest duration; it can last minutes. Large power transformers may become inoperable due to the electric currents produced along power lines by Component 3.

1.3 Nuclear Energy

Two types of nuclear reactions can provide nuclear energy: nuclear fission, and nuclear fusion. Nuclear fission occurs when a large nucleus splits into smaller nuclei and releases energy. The fission process was used to prepare the first nuclear weapons. Nuclear fusion, by contrast, occurs when two nuclei combine, or fuse, to form one larger nucleus with the release of energy. Energy from the sun is provided by nuclear fusion reactions.

The release of energy from a nuclear reaction needs to be controlled. In cogeneration, some waste heat from the release of energy is captured and transformed into a more useful form. Decay products of a fission process can remain highly radioactive for a very long time and must be disposed of in a way that does not harm the environment. Byproducts of the fusion process, however, are generally safe.

Unlike nuclear weapons, which are designed to provide an explosive release of nuclear energy, nuclear reactors are designed to control the release of nuclear energy. The only commercially viable nuclear reactors are based on nuclear fission. The fusion process is not currently available in a commercial nuclear reactor.

The first commercial nuclear fission power plant began operation in 1957 and output 60 MW of electric power (Murray, 2001, p. 202). The power plant was built at Shippingport, Pennsylvania on the Ohio River. Shippingport was a city approximately 25 miles from Pittsburgh, Pennsylvania. Today, several nations, including the United States, China, France, and Russia, generate a significant percentage of their electricity using nuclear fission power plants.

Fusion

The basic principle of nuclear fusion is to release vast amounts of energy when two nuclei are fused to form a larger nucleus. Modern nuclear fusion technology relies on collisions between protons, deuterium nuclei (deuterons), and tritium nuclei (tritons). Figure 1.5 illustrates the fusion of deuterium and tritium to form helium, a neutron, and excess energy.

The materials needed for fusion reactions include protons, deuterium, and tritium. Protons are available as hydrogen nuclei. Deuterium is an isotope of hydrogen and has a nucleus with one proton and one neutron. Ordinary water contains around 0.015 mole percent deuterium, or one deuterium atom for every 6700 hydrogen atoms (Murray, 2001, p. 77). Tritium is a radioactive isotope of hydrogen with one proton and two neutrons in its nucleus. Tritium can be produced in fission reactor cores by the irradiation of lithium rods by excess neutrons.

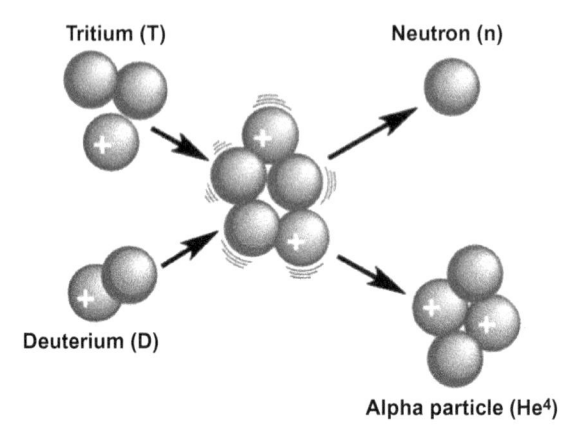

Figure 1.5 Illustration of a Fusion Reaction $D + T \rightarrow He^4 + n$

Figure 1.6 ITER Facility (Photo Courtesy of ITER Organization/EJF Riche, March 16, 2023 (ITER, 2023))

The production of commercial nuclear fusion power sometime in the 21st century is the aim of nuclear fusion research and development. Due to the technical challenges involved in starting and managing a fusion reaction, attempts to commercialize fusion energy have not yet been successful. Commercialization of nuclear fusion energy depends on controlling the process.

The International Thermonuclear Experimental Reactor (ITER), depicted in Figure 1.6, is being built in Southern France (ITER, 2023). The ITER project is sponsored by China, the European Union, India, Japan, Korea, Russia, and the United States. The goal of ITER is to demonstrate the commercial-scale viability of nuclear fusion reactor technology. Deuterium and tritium will be used as nuclear fuel. The byproduct of deuterium-tritium fusion is helium, an environmentally safe inert gas. Fusion energy is anticipated to make a significant contribution to the global energy mix by the end of the 21st century. More information about nuclear fusion is presented by Fanchi (2024).

Chapter 2

What Paths Are Possible?

Several key players and their historical importance to the emergence of the modern world order were discussed in Chapter 1. Here, we use historical analysis to identify paths that could be followed as the world order develops.

2.1 Evolution and Lifetime of Empires

Historians have recognized a pattern in the evolution and lifetime of empires. Three theories about the fate of empires are outlined in this chapter. The first theory is attributed to John Bagot Glubb. He was a British general and historian who fought in World War I for Belgium and France, then spent much of his career in the Middle East. The second theory was proposed to explain the rise and fall of democracies. It is summarized by the Prentis Cycle described in Section 2.1.2. The third theory is Alexis de Tocqueville's study of democracy in America.

2.1.1 *The Fate of Empires*

John Bagot Glubb developed a theory of the fate of empires by studying several great states (Glubb, 1976). Table 2.1 summarizes Glubb's estimates of the length of time it took several great states to rise and then fall. Over the course of 3000 years of human history, a great state typically lasted about 250 years, or approximately 10–12 generations.

Table 2.1 Estimated Lifetimes of Great States

State	Dates of Rise and Fall	Duration (years)
Assyria	859–612 BC	247
Persia (Cyrus and descendants)	538–330 BC	208
Greece (Alexander and successors)	331–100 BC	231
Roman Republic	260–27 BC	233
Roman Empire	27 BC–180 AD	207
Arab Empire	634–880	246
Mameluke Empire	1250–1517	267
Ottoman Empire	1320–1570	250
Spain	1500–1750	250
Romanov Russia	1682–1916	234
Britain	1700–1950	250

Table 2.2 Stages in the Rise and Fall of Great States

#	Stage
1	The Age of Pioneers (outburst)
2	The Age of Conquests
3	The Age of Commerce
4	The Age of Affluence
5	The Age of Intellect
6	The Age of Decadence

In addition to estimating the lifetime of great states, Glubb observed a pattern in the histories of the empires he studied. First, the state was established, then grew, matured, declined, and finally collapsed. Table 2.2 lists the stages (Glubb, 1976).

Glubb believed that the last age, the Age of Decadence, resulted from a lengthy period of prosperity and power; selfishness; a love of money;

Table 2.3 Characteristics of Decadence

No.	Characteristic
1	Defensiveness
2	Pessimism
3	Materialism
4	Frivolity
5	An influx of foreigners
6	The Welfare State
7	A weakening of religion

and a loss of social responsibility. The loss of social responsibility can also be considered a loss of patriotism or loss of a sense of duty to the state. The characteristics of decadence are listed in Table 2.3 (Glubb, 1976).

Glubb believed that the life cycles of great states were impacted by similar internal events up to the Age of Decadence. The lifetimes of great states varied because the Age of Decadence was affected by external factors that varied from state to state. As an example, one characteristic that contributes to the Age of Decadence is the influx of foreigners. The unity that held the initial state together may be weakened by the shifting demographics of the core population that formed and expanded the state.

2.1.2 *The Prentis Cycle*

Henning W. Prentis, Jr. delivered an address on democracy called the "Cult of Competency" at the Mid-Year Convocation of the University of Pennsylvania in the middle of World War II (Prentis, 1943). Prentis noted that "The Greeks said, Know yourself; the Romans, Be yourself; the Christians, Give yourself. The cult of competency — as our fathers knew and practiced it — fused these three swords of the spirit into the great instrumentality that set them free."

He reminded his audience that freedom "must be won and re-won by every generation for itself."

Prentis suggested that democracies have been subject to the following historical cycle: "From bondage to spiritual faith; from spiritual faith to courage; from courage to liberty; from liberty to abundance, from abundance to selfishness; from selfishness to apathy, from apathy to dependency; and from dependency back to bondage once more." This cycle sequence is known as the Prentis Cycle. It is illustrated in Figure 2.1. This cycle is often incorrectly attributed to Scottish historian Alexander Fraser Tytler (1747–1813) and called the Tytler Cycle (Collins, 2009).

One possible justification of the Prentis Cycle begins with a community in bondage. Such a community has few, if any, freedoms. The community turns to religion and religious faith for comfort and courage. Members of the community need courage to fight for and earn their freedom. Once the community is liberated, it is free to produce an abundance of material goods.

The second half of the cycle or loop describes the decline of the community. Selfish consumption of goods can lead to complacency and eventually apathy. The community becomes dependent on others and returns to bondage.

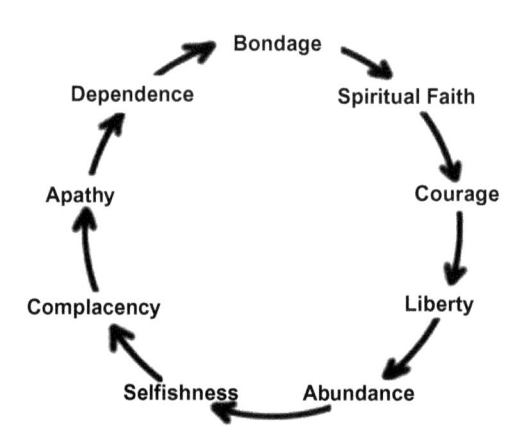

Figure 2.1 The Historical Prentis Cycle (Prentis, 1943)

Point to Ponder: Ronald Reagan and Freedom

Ronald Reagan was inaugurated as President of the United States in 1981. Earlier in his political career, Reagan pointed out the fragility of freedom: "Freedom is never more than one generation away from extinction. We didn't pass it on to our children in the bloodstream. The only way they can inherit the freedom we have known is if we fight for it, protect it, defend it, and then hand it to them with the well-fought lessons of how they in their lifetime must do the same. And if you and I don't do this, then you and I may well spend our sunset years telling our children and our children's children what it once was like in America when men were free" (Reagan, 1961). Reagan used variants of this statement many times during the remainder of his career.

2.1.3 *de Tocqueville's Tyranny of the Majority*

Alexis de Tocqueville, a 25-year-old French aristocrat, visited the United States in 1831, during the Andrew Jackson presidency. de Tocqueville's official function was to examine the penal system in the fledgling nation, but he was also interested in developing a better understanding of America and American democracy. He traveled for nine months, usually by steamboat but also on foot and by horseback. He visited Eastern cities, journeyed along the Ohio and Mississippi rivers, and explored the northwestern wilderness. He met Americans of every rank and occupation. de Tocqueville fulfilled his official function by writing a book about the American penal system, but his journey is best remembered for his observations on sovereignty and freedom that he recorded in *Democracy in America* (de Tocqueville, 1966).

de Tocqueville's *Democracy in America* was released in two volumes in 1835 and 1840. Volume One discussed the physical configuration of America, its jurisdictions, and political institutions. Volume Two investigated the impact of American democracy on what are now considered cultural topics, such as economics, literature, religion, and the family.

In the Introduction in Volume 1, de Tocqueville outlined his belief that the world was moving toward democracy. The American Revolution began in 1776 and the French Revolution began in 1789. Both revolutions overthrew monarchies and foreshadowed the formation of more democratic systems of governance. de Tocqueville believed that by studying America, we could better understand democracy and glimpse the future of the world: "I admit," he said, "that I saw in America more than America; it was the shape of democracy I sought, its inclinations, character, prejudices, and passions; I wanted to understand it so as at least to know what we have to fear or hope therefrom" (de Tocqueville, 1966, Volume 1, p. 12).

de Tocqueville believed that public opinion could lead to a tyranny of the majority in a democracy, that is, a majority of people might choose to impose their will on marginalized people or other minorities of people with different views. He explains his position in Volume 1, Part 2, Chapter 7, "The Omnipotence of the Majority in the United States and Its Effects." He recognized that "The absolute sovereignty of the will of the majority is the essence of democratic government, for in democracies there is nothing outside the majority capable of resisting it" (de Tocqueville, 1966, Volume 1, p. 227). He said that, in the United States, "all the parties are ready to recognize the rights of the majority because they all hope one day to profit themselves by them.

Hence the majority in the United States has immense actual power and a power of opinion which is almost as great. When once its mind is made up on any question, there are, so to say, no obstacles which can retard, much less halt, its progress and give it time to hear the wails of those it crushes as it passes" (de Tocqueville, 1966, Volume 1, p. 229).

After acknowledging the power of the majority, de Tocqueville said that "when I refuse to obey an unjust law, I by no means deny the majority's right to give orders; I only appeal from the sovereignty of the people to the sovereignty of the human race" (de Tocqueville, 1966, Volume 1, p. 231). He concluded that "What I find most repulsive in America is not the extreme freedom reigning there but the shortage of guarantees against tyranny" (de Tocqueville, 1966, Volume 1, p. 233).

Point to Ponder: Raiding the Treasury

An example of the tyranny of the majority can be expressed by the following statement: 'A democracy will continue to exist up until the time that voters discover they can vote themselves generous gifts from the public treasury. From that moment on, the majority always votes for the candidates who promise the most benefits from the public treasury, with the result that every democracy will finally collapse due to loose fiscal policy, which is always followed by a dictatorship.' This statement is often presented as a quote and attributed to either de Tocqueville or Alexander Fraser Tytler. It appears that the source of the statement is unknown, even though the statement captures a sentiment that many authors choose to discuss.

2.2 Clash of Civilizations

Samuel P. Huntington, an American political scientist, predicted that conflicts in the future would more likely be fought between cultures than between nations (Huntington, 1996). His view contributes to our understanding of sociopolitical events that influence energy consumption and the emergence of a new world order.

According to Huntington, a geopolitical paradigm represented the presumptions, ideas, standards, and customs that formed the basis of the world order before World War II. The paradigm could undergo a shift, or change, following one or more major events. This pre-war paradigm helped people comprehend and anticipate events before World War II but was no longer applicable after the war. Post-war events, such as the Cold War between the United States and the Soviet Union, were leading to a paradigm shift, or change, in paradigm.

The framework created during the Cold War between the Soviet Union and the Western alliance led by the United States helped people understand international affairs after World War II. The end of the Cold War was marked by the fall of the Berlin Wall and the break-up of the Soviet Union in the late 1980s and early 1990s. A new geopolitical paradigm was needed.

Table 2.4 Geopolitical Paradigms Proposed by Huntington

1. One Unified World
2. Two Worlds
3. Anarchy (approximately 195 nation-states)
4. Chaos

Several geopolitical models are represented by paradigms Huntington proposed. They are listed in Table 2.4 (Huntington, 1996). The One Unified World paradigm models the world as relatively unified under a supranational government during the comparatively peaceful period following the end of world wars. The second paradigm, the Two Worlds paradigm, models a world that is geopolitically polarized. During the Cold War, nations typically aligned with either the United States or the Soviet Union. Today, nations align with either democracies led by NATO or totalitarian regimes led by China and Russia. Non-aligned states have been present since World War II and represent a third, non-aligned "world" that can influence events. A third paradigm, Anarchy, models nations as independent actors that pursue their own interests.

A fourth paradigm, the Chaos paradigm, recognizes that alliances between states are dynamic. Allegiances are no longer between nations but between people based on common traditions and value systems in a world where information is readily available. New allegiances based on such factors as ancestry, language, and religion are forming alliances that can result in civilizations that can change with time. Table 2.5 presents Huntington's list of civilizations. The list is subject to revision.

Huntington considered the Chaos paradigm the most accurate representation of historical events at the end of the 20th century. The Chaos paradigm helps explain relationships between important contemporary civilizations. The following examples illustrate the application of Huntington's ideas.

The Role of Core States

Huntington identified at least one core state associated with each civilization (Huntington, 1996, Chapter 7). Core states of the Western Civilization include France and Germany of the European Union, and the United States.

Table 2.5 Huntington's Major Contemporary Civilizations

Civilization	Comments
Sinic	China and related cultures in Southeast Asia.
Japanese	The distinct civilization that emerged from the Chinese civilization between 100 and 400 A.D.
Hindu	The peoples of the Indian subcontinent that share a Hindu heritage.
Islamic	A civilization that originated in the Arabian Peninsula and now includes subcultures in Arabia, Turkey, Persia, and Malaysia.
Western	A civilization centered around the northern Atlantic that has a European heritage and includes people in Europe, North America, Australia, and New Zealand.
Orthodox	A civilization centered in Russia and distinguished from Western Civilization by its cultural heritage, including limited exposure to Western experiences (such as the Renaissance, the Reformation, and the Enlightenment).
Latin America	People with a European and Roman Catholic heritage who have lived in authoritarian cultures in Mexico, Central America, and South America.

Russia is a core state in Orthodox Civilization, and China is a core state in Sinic Civilization. An alliance between two strong core states, like China and Russia, can lead to the polarization of civilizations.

Western Civilization is an example of a civilization where stable relations between core states within the civilization have enhanced the stability of the civilization. On the other hand, one of the main causes of unrest within the Islamic Civilization appears to be the absence of stable relationships between its core states. The Organization of Petroleum Exporting Countries (OPEC) has served several Islamic states since its inception in 1960 and could help stabilize core state relationships.

Impact on Multiculturalism

Multicultural states have seen the growth of communities within their borders that may not share the values and allegiances of the host state.

Many European countries, for example, are members of Western Civilization with sizable Islamic populations. As cultures within member states compete for dominance, the rise of multiculturalism and sizable immigrant populations can change the cultural identities of some states within a civilization. A shift in cultural identification may result in a change in civilizational loyalty.

Some multicultural states are held together by force rather than common interests. For example, Yugoslavia and Czechoslovakia were bound together within the Soviet Union by powerful central governments. The dissolution of the Soviet Union was followed by the dissolution of the multicultural states into separate states with more compatible values.

Global migration affects the multicultural status of nations. Many Muslims from the Middle East migrated to Europe and the United States in 2015 because of instability in member states of Islamic Civilization. Today, the Biden Administration's refugee policy, which went into effect in 2021, has caused a significant increase in immigration to the United States. If the migration is large enough, it is possible to alter the cultural identity of the host country. The likelihood of an undesirable cultural shift can worry residents of host countries and raises the question of whether the migration is an act of goodwill by the host country or an invasion by a different civilization. Glubb identified an influx of foreigners as a characteristic of the Age of Decadence (Table 2.3).

Conflicts within Civilizations

World Wars I and II in the 20th century demonstrate that states within a civilization do not always live in harmony. Differences between states can result in conflicts within a civilization that can lead to civil wars. From this point of view, World Wars I and II can be considered civil wars in Western Civilization. The civil wars expanded to engulf states from other civilizations.

Conflicts between Civilizations

Conflicts between civilizations are exemplified by the Cold War and oil crises. For centuries, Western civilization has been the most powerful civilization in the world, where "power" is defined as the capacity to manage and influence the actions of others. An analysis of the relative military power and energy dependence of key states reveals that the influence of Western Civilization is waning in world politics as other civilizations advance technologically and economically. Evidence for this observation can be seen in the relationship between Western and Islamic civilizations.

Today, oil is ordinarily exported by oil-producing member states of the Islamic Civilization to member states of the Western Civilization. As a result, wealth moves to oil-exporting states in Islamic Civilization from oil-importing states in Western Civilization. Most member states of the Islamic Civilization lack nuclear weapons and have relatively weak militaries when compared with core states in the Western Civilization, such as the United States. The acquisition of wealth by member states of the Islamic Civilization allows them to change the balance of power between Islamic Civilization and Western Civilization.

Iran is one of the core states of the Islamic Civilization. It is developing a nuclear capability that could be used for either peaceful or military applications. Other Islamic nations without nuclear weaponry are worried about the potential acquisition of nuclear weapons by Iran. For example, Saudi Arabia and Iran (formerly Persia) are both Islamic countries and members of OPEC but have different religious views within the Islamic faith. In a world where increased use of renewable energy sources and shale oil production are reducing the demand for imported oil, rivalry for market share is a factor that strains relations between OPEC nations, many of which are members of Islamic Civilization.

2.3 Clash over Resources

Core states often seek to spread civilizational differences by extending their global influence. The clash of civilizations may be fought on economic, ideological, or even military fronts. Access to key resources such as energy can impact the outcome of these clashes.

States in a civilization that import energy are dependent on states in other civilizations that export energy. Energy becomes a weapon in a conflict between civilizations when relations between states that trade in energy are hostile. For example, population growth in civilizations such as the Sinic and Hindu Civilizations has increased demand for a limited supply of oil to increase quality of life. The increase in demand for oil without an increase in supply leads to an increase in oil prices and an increased transfer of wealth from oil-importing states to oil-exporting states.

States that rely on imported oil may try to reduce demand by implementing conservation measures or developing alternative energy sources. Environmentalists and energy conservationists are political forces in some nations, such as the Green Party in Germany and the Sierra Club in the United States. It is not unusual for people to support energy conservation initiatives until they are confronted with the cost or inconvenience of implementing energy-saving activities like recycling, carpooling, or purchasing energy-efficient automobiles.

Access to natural resources depends on the type of relationships that exist between states with the technology to produce natural resources and states with territorial control over those resources. Oil production is a good example of this observation.

Oil production technology was primarily developed by member states in Western Civilization. Access to technology and adequate oil resources within its borders provided Western Civilization with the energy required to grow into the most powerful civilization in history. Oil resources are limited and non-renewable, however. The consumption of available oil resources led member states of Western Civilization to seek alternative sources. Western Civilization has become dependent on other civilizations for the oil it needs to support its oil-dependent economies. Unfortunately, the relationships between oil-importing states and oil-exporting states are not always cooperative and can cause conflict between civilizations.

Oil-importing states are developing alternatives to fossil fuels as a means of minimizing their dependence on energy resources from oil-exporting states. The European Union (EU) has been pursuing a program to achieve energy independence by converting member state economies to renewable energy sources by 2050. The Russian invasion of Ukraine that began in 2022 has forced EU decision-makers to reevaluate their plans and alter their timelines.

Huntington provided one set of possible paradigms, such as those in Table 2.4, to help explain contemporary geopolitical events. Klare (2004) provided an alternate point of view: competition for limited resources is what drives contemporary geopolitics. According to Klare (2004, p. xii): "After examining a number of recent wars in Africa and Asia, I came to a conclusion radically different from Huntington's: that resources, not differences in civilizations or identities, are at the root of most contemporary conflicts."

Klare argued in his book *Resource Wars* (Klare, 2001) that oil, water, land, and minerals were all significant enough to serve as sources of conflict, but in his book *Blood and Oil* (Klare, 2004), Klare focused on petroleum. Klare observed that modern economies are still dependent on petroleum which is still relatively cheap and plentiful. Efforts to abandon fossil fuels as soon as possible, including petroleum, are motivated by climate change concerns rather than supply concerns (Fanchi, 2024).

Daniel Yergin, an energy expert, has observed that the world map is being redrawn by nations seeking access to the resources needed to fuel modern and emerging economies (Yergin, 2020). Another energy expert, Alex Epstein, presents a moral case that oil is needed to raise the quality of life in developing countries (Epstein, 2014, 2022).

Part 2
Weaponizing Environmentalism

Chapter 3
Environmental Socialism

The modern world order is being affected by the energy transition. Part 1 explained how the modern world order emerged and described different factors that can catalyze changes to the existing world order. In Parts 2–4, we concentrate on the main mechanisms being employed to shape a new world order under a global government following the energy transition. In Part 2, we explain how the environment can be weaponized to help establish a global government.

The historical emergence of a global government in the modern era is described in Chapters 3–5. The League of Nations created after World War I and the United Nations created after World War II are examples of multinational organizations that were founded as global governments. In Chapter 6, we explain why U Thant, the first third-world Secretary General of the United Nations, chose to use the environment as a tool for exercising political influence over sovereign states.

3.1 Global Governance

It is important to recognize that global governance is not the same as global government (Haass, 2020, p. xxi). Global governance does not require the existence of a single ruling entity with absolute power. Instead, it refers to international cooperation among institutions that coordinate the behavior of transnational players, mediate disputes, and mitigate issues arising from collective action. In general, global governance entails making, monitoring, and enforcing rules for transnational activities.

By contrast, global government, also known as world government, refers to a single political authority with jurisdiction over all of humanity. It would have the power to overrule the governments of sovereign nations and all other subordinate levels of government. International organizations like the League of Nations and the United Nations are examples of attempts to establish institutions with global authority. We review their founding below and discuss the role of the Cold War in helping motivate the weaponization of the environment by the United Nations, a relatively weak but emerging global government.

3.2 The Political Roots of Socialist Environmentalism

The implementation of energy transition policies can be impeded by sovereign states and geopolitical relationships. For example, the Cold War between the United States and the Soviet Union sparked a struggle for influence on a worldwide scale. Both the United States and the Soviet Union were often forced to engage in military confrontations and diplomatic initiatives to preserve their positions as global superpowers. The global danger emanating from the Cold War and the desire to implement a socialist global government motivated the development of socialist environmentalism. The path to socialist environmentalism begins with the question: Is modern environmentalism an attempt to impose socialism?

Brian Sussman argued that contemporary environmentalism is an attempt to impose socialism worldwide (Sussman, 2012). He traced the origins of the movement back to Karl Marx (1818–1883), an advocate of organized collectivism, commonly known as socialism or communism.

Marx was influenced by Georg Hegel's (1770–1831) philosophy while Marx was a student at the University of Berlin. Hegel believed that worldviews based on religious beliefs and faith, such as Judaism and Christianity, should be replaced by a new worldview based on scientific reason and truth.

Jewish and Christian faiths believe that possessions have value, and people have a role to play in history (Smith, 1991, Chapter VII). The concept that God created a material world is expressed in the first sentence of Genesis, the first book of the Old Testament: "In the beginning, God created heaven and earth" (Torah, 1992, Genesis 1:1; see also RSV Bible, 1971, Genesis 1:1). Genesis also says that "God saw all that He had made and found it very good" (Torah, 1992, Genesis 1:31; see also RSV Bible,

1971, Genesis 1:31). The idea that God created the material world and said that it is good assigns value to the existence of matter. In addition, Genesis implies that God supports the ownership of property when it says that humans should "fill the earth and master it; and have rule over the fish of the sea, the birds of the sky, and all the living things that creep on earth" (Torah, 1992, Genesis 1:28; see also RSV Bible, 1971, Genesis 1:28).

The Genesis account of creation says that the entry of man into the material world was predetermined. Furthermore, Judaism teaches that God decrees a purpose to historical events (Smith, 1991, Chapter VII). Consequently, human activities, institutions, and social order are endowed with value because they can influence the course of history by either helping or hindering the accomplishment of God's purpose.

Marx and his collaborator Friedrich Engels (1820–1895) were interested in property and humanity's role in history. They saw a connection between dialectics, materialism, and religion. Plato (ca. 428–348 B.C.) used the term dialectic in the Socratic dialogues. Hegel introduced a dialectical process that relied on reason to resolve disagreements between opposing points of view. Dialectic differs from debate because debate can utilize any tactic to persuade, whereas dialectic must use reasoned argument. Hegel represented differing viewpoints as a thesis and its antithesis. He then employed dialectical reasoning to seek a synthesis of the differing viewpoints. Marx and Engels developed dialectical materialism in the 1920s by combining concepts from materialism and dialectics.

3.3 Dialectical Materialism

In 1841, Karl Marx completed his doctoral dissertation in philosophy (Marx, 1841) entitled "The Difference Between Democritean and Epicurean Philosophy of Nature." Epicurus, a classical Greek philosopher, presented a philosophy of matter that was known in the 1800s as materialism. The Epicurean view of the world says that the world consists only of matter, which consists of indivisible particles of matter called atoms that are too small to see.

Marx and Engels shared the Epicurean view of the material world. Engels wrote that "if science can get to know all there is to know about matter, we will then know all there is to know about everything" (Engels, 1886). Many modern researchers agree with a materialistic view of the world (Kragh, 2004).

Table 3.1 Laws of Matter Proposed by Marx and Engels

No.	Law of	Description
1	Opposition	Nature contains objects with opposing characteristics
2	Negation	Species can proliferate by negating themselves or dying
3	Transformation	Species can change (evolve)

Dialectical materialism combines the ideas of dialectics with materialism. It says that everything in the world is composed of matter, and a material system can achieve a more stable state by settling the conflict between conflicting attributes.

Table 3.1 summarizes three Laws of Matter proposed by Engels: the Law of Opposition, the Law of Negation, and the Law of Transformation. The Law of Opposition says that objects with opposing attributes can co-exist when they achieve a state of equilibrium. For example, conflict between humans can arise when humans do not share compatible values. In this view, government is often called upon to resolve human conflict.

The Law of Negation makes it possible for a species to reproduce in larger numbers because a preceding generation made way for a succeeding generation. The current population explosion demonstrates that humans struggle with self-regulation of their reproduction. Consequently, government intervention may be necessary to achieve sustainability.

The Law of Transformation says that the change, or evolution, of a species, should be viewed as progressive, that is, the evolution of a species seeks to improve the species. Consequently, one segment of the species may make a significant evolutionary leap while coexisting with a less evolved segment of the species. According to the concept of organized collectivism, the segment of humanity with elite status should serve as the ruling class and take care of the species.

The three Laws of Matter, according to Sussman, "provide the rationale for today's green agenda" (Sussman, 2012, p. 3). Furthermore, a society that emerges from the three Laws of Matter is expected to reject absolutism — the belief in absolute truth — and welcome relativism. Such a society will be dependent on humanity and reject the supernatural.

3.4 The Marxist View of Property

The American Declaration of Independence (1776) stated that "We hold these truths to be self-evident, that all men are created equal, that they are endowed by their Creator with certain unalienable Rights, that among these are Life, Liberty and the pursuit of Happiness" (Founders, 1776). Hegel and Marx rejected these fundamental ideas.

Hegel said that "The state as a completed reality is the ethical whole and the actualization of freedom. It is the absolute purpose of reason that freedom should be actualized" (Hegel, 1820, p. 197). Consequently, Hegel believed that the state was responsible for granting freedom. Likewise, Marx believed that the life of the individual had value only within the collective. Marx believed that "philosophers have only interpreted the world, in various ways; the point, however, is to change it" (Marx, 1845, Thesis XI). Marx tried to facilitate change on a global scale. As an atheist that believed in materialism and the collective, Marx did not accept the beliefs that individuals have rights endowed by God, such as rights to life, liberty, and the pursuit of happiness.

The writings of political philosopher John Locke (1632–1704) influenced the American Declaration of Independence. Locke wrote that the "necessity of pursuing happiness [is] the foundation of liberty" as a justification for private property ownership (Locke, 1690, Chapter XXI, Paragraph 52). He attempted "to show how men might come to have a property in several parts of that which God gave to mankind in common" (Locke, 1680, paragraph 25, Chapter 5 entitled "Of Property," 2nd treatise). According to Locke, "every man has a 'property' in his own 'person.' ... The 'labour' of his body and the 'work' of his hands are properly his. Whatsoever, then, he removes out of the state that Nature hath provided and left it in, he hath mixed his labour with it, and joined to it something that is his own, and thereby makes it his property" (Locke, 1680, Paragraph 27, Chapter 5, 2nd treatise).

Property can include material assets like real estate, goods, or money. It can also refer to intangibles like the right of an individual to express personal thoughts and opinions. James Madison, a founder of the American Republic, wrote a 1792 Essay entitled Property. He attempted to explain the meaning of property by stating that property "embraces everything to which a man may attach a value and have a right; and which leaves to everyone else like advantage" (Madison, 1792).

The means of production are owned and controlled by the bourgeoisie, whereas the proletariat is comprised of workers. Marx saw capitalism as a struggle between the bourgeoisie and the proletariat classes. According to Marx and Engels, "The distinguishing feature of Communism is not the abolition of property generally, but the abolition of bourgeois property. Modern bourgeois private property is the final and most complete expression of the system of producing and appropriating products that is based on class antagonisms, on the exploitation of the many by the few.

"In this sense, the theory of the Communists may be summed up in the single sentence: Abolition of private property" (Marx and Engels, 1848, Section II, p. 223).

3.5 The Marxist View of Natural Resources

Sussman cites German chemist Justus Liebig (1803–1873), a pioneer in organic chemistry, as the earliest example of a scientist opposing capitalism because of its impact on the environment (Sussman, 2012, p. 8). Liebig noted that capitalist agriculturalists used guano (bird droppings) as fertilizer. It had an unforeseen and adverse environmental impact.

Liebig explained that the "barren soil on the coast of Peru is rendered fertile by means of a manure called Guano, which is collected from several islands on the South Sea" (Liebig, 1840, p. 74). He said that guano "forms a stratum several feet in thickness upon the surface of these islands, (and) consists of the putrid excrements of innumerable seafowl that remain on them during the breeding season" (Liebig, 1840, p. 74, footnote). Phosphate-rich guano was a significant agricultural ingredient in many regions of the world, including Europe and the Americas, because of its high nutritional value.

The widespread use of guano concerned Liebig. He believed that a society should replace all the nutrients it took out of the earth, otherwise the agricultural system was illegitimate. He believed "the commercial farming system of contemporary Europe violated this principle" (Märald, 2002, p. 73).

As an illustration, Liebig noted that Great Britain, a leader in European agriculture, "deprives all countries of the conditions of their fertility. It has raked up the battlefields of Leipsic, Waterloo, and the Crimea; it has consumed the bones of many generations accumulated in the catacombs of Sicily; and now annually destroys the food for a future generation of 3 millions and a half of people. Like a vampire it hangs on the breast of Europe, and even the world, sucking its lifeblood without any real necessity or permanent gain for itself" (Märald, 2002, p. 74).

Marx's understanding of ground rent was influenced by Liebig's work. Ground rent gives the tenant the right to use a plot of land in exchange for payment to a landowner. The ground rent agreement allows the tenant to own property on or improvements to the land. Marx stated in a letter to Engels that he "had to plough through the new agricultural chemistry in Germany, in particular Liebig and Schönbein, which is more important for this matter [ground rent] than all the economists put together" (Marx, 1866). Marx believed that one of Liebig's achievements was to have "developed from the point of view of natural science, the negative, i.e., the destructive side of modern agriculture" (Marx, 1887, p. 357, note 246).

According to Marx, "all progress in capitalistic agriculture is a progress in the art, not only of robbing the laborer, but of robbing the soil; all progress in increasing the fertility of the soil for a given time, is a progress towards ruining the lasting sources of that fertility" (Marx, 1887, p. 330). He concluded that the "moral of history is that the capitalist system works against a rational agriculture, or that a rational agriculture is incompatible with the capitalist system" (Marx, 1894, p. 83).

3.6 The Marxist View of Environmentalism Goes National

Foster traced a linkage between dialectical materialism, Karl Marx, Charles Darwin, and science in Britain (Foster, 2002, p. 81). This linkage began with British zoologist Edwin Ray Lankester (1847–1929). Lankester became a "close friend of Karl Marx in the last few years of Marx's life" (Foster, 2000) while Lankester was a professor at University College, London. Lankester was a Darwinist and protégé of Thomas Henry Huxley (1825–1895) (Milner, 1999, p. 90; Foster, 2002, p. 82). Huxley was a prominent supporter of Charles Darwin's (1809–1882) work on natural selection and was acquainted with Lankester through his friendship with Lankester's father.

Lankester agreed with Marx and Liebig that humans were having an adverse impact on nature. He pointed out in *The Effacement of Nature by Man* that man's activities were responsible for "vast destruction and defacement of the living world by the uncalculating reckless procedure of both savage and civilized man" (Lankester, 1913, p. 365).

Lankester concluded that "so far as we can see, if man continues to act in the reckless way which has characterized his behavior hitherto, he will multiply to such an enormous extent that only a few kinds of animals will be left on the face of the globe" (Lankester, 1913, p. 366).

Lankester and Arthur Tansley (1871–1955) were among the socialists who infused naturalistic and ecological principles into their way of thinking (Foster, 2002, pp. 79–80). While studying at University College London, Tansley was influenced by Lankester and botanist Francis Wall Olivier. Lankester and Tansley were Fabian socialists rather than Marxist socialists. We begin our discussion of Fabian socialism here by introducing the Fabian Society, and then consider Fabian socialism in more detail in the following chapter.

The Fabian Society is a British socialist organization founded in 1884 in London. Its goal was to gradually bring about a collectivist one-world government by gradually reforming democracies. Fabian Socialists hope to advance the ideals of democratic socialism. Fabian Socialists believe that a gradual transition to democratic socialism would be more effective than imposing Marxist doctrine by revolutionary upheaval. Fabian Socialism would achieve a global socialist state by gradually reforming democratic systems.

Lankester and Tansley shared a materialist philosophy. Tansley is credited with creating the word "ecosystem." He thought that people were damaging the environment by their actions. Tansley (1900–1991) was a colleague of Charles Elton at the Nature Conservancy (Bocking, 2012). Elton applied his "fiery" writing style to criticize pesticide use in 1958 (Elton, 1958, pp. 137–142).

Vladimir Lenin (1870–1924) was aware that Marx depended on Liebig's agricultural research (Lenin, 1901). Lenin belonged to the Bolshevik Party, a faction of the Russian Social-Democratic Workers Party. In 1917, Lenin led the Bolshevik Revolution and became the first Soviet leader. In 1918, the Bolshevik Party was renamed the Russian Communist Party.

Lenin worried that private ownership of agriculture in a capitalist economy could damage the environment. He issued two decrees as head of the Soviet state in October 1917. The first decree withdrew Russia from World War I, and the second decree entitled "On Land" gave the government control over natural resources such as land, minerals, waterways, and forests.

Lenin's Central Executive Committee passed legislation in 1918 that codified his second decree. It was entitled "The Fundamental Law of Land Socialization." The legislation abolished private ownership of land in Part 1, Article 1 by asserting that "All private ownership of land, minerals, waters, forests, and natural resources within the boundaries of the Russian Federated Soviet Republic is abolished forever" (Lenin CEC, 1918). Part 1, Article 2 of the legislation required that all land be "handed over without compensation (open or secret) to the toiling masses for their use" (Lenin CEC, 1918). Part 1, Article 17 imposed Lenin's belief that natural resources should not be used for profit. It stated that "Surplus income derived from the natural fertility of the soil or from nearness to market is to be turned over to the organs of the Soviet Government, which will use it for the good of society" (Lenin CEC, 1918).

Chapter 4
Fabian Socialism

Karl Marx and Friedrich Engels believed that the proletariat would need to rebel against the bourgeoisie control of the government if the proletariat was to seize power. Vladimir Lenin led a revolution in 1917 to replace proletariat rule under Czar Nicholas II with a Marxist system.

The view expressed by Marx and Engels that a revolution would be required to replace bourgeoisie rule with proletariat rule was not universally accepted. One group, the Fabian Society, believed that stable democracies could be replaced with Marxist or socialist governments by gradually replacing a democratic system with a Marxist or socialist system. Rather than imposing revolutionary change, members of the Fabian Society would pursue a strategy of evolutionary change, that is, a strategy based on "the belief that radical long-term goals are best advanced through empirical, practical, gradual reform" (Fabian Society, 2023). The result of evolutionary change would be a progressive transition to a new government rather than an abrupt transition. The Fabian approach to socialism is discussed here.

4.1 The Fabian Society

The Fabian Society was named after the Roman general Quintus Fabius Maximus Verrucosus (ca. 280–203 B.C.). Verrucosus earned the nickname Cunctator, or Lingerer when he chose to delay attacks on Hannibal's invading Carthaginian army during the Second Punic War. According to the first Fabian pamphlet: "For the right moment you must wait, as Fabius did most patiently when warring against Hannibal, though many censured his delays; but when the time comes you must strike hard, as Fabius did, or your waiting will be in vain, and fruitless" (Pease, 1916, Chapter III).

Thomas Davidson (1840–1900) was a Scottish-American philosopher and lecturer. He believed that social reform begins with the education of the individual. According to Edward Pease, who was a founding member of the Fabian Society and served as secretary and historian, Davidson founded Vita Nuova, or the Fellowship of the New Life, in 1883 London (Pease, 1916) and subsequently in New York. In 1884, Davidson began the Fabian Society as an offshoot of the Fellowship of the New Life. Pease wrote that Davidson "was the occasion rather than the cause of the founding of the Fabian Society. His socialism was ethical and individual rather than economic and political" (Pease, 1916, Chapter 1).

In a tribute to Davidson, Charles Blakewell explained Davidson's motivation for founding the Fabian Society: "the times were religiously and socially out of joint, and (Davidson believed) that it was his duty, as it was that of every man, to do his best to set them right. With this end, he took an active interest in the founding of the London Fabian Society" (Blakewell, 1901, p. 447). Blakewell noted that the founding goals of the Fabian Society were to gather and spread knowledge that could improve social conditions. He noted that Davidson's interest in the Fabian Society waned as the organization drifted toward socialism.

Edward Pease, Edith Nesbith, Hubert Bland, and Frank Podmore were founding members of the Fabian Society. A few notable early members of the group included Irish playwright George Bernard Shaw (1856–1950), Sydney Webb (husband of author Beatrix Potter who later joined the Fabian Society), and Eleanor Marx, the eldest daughter of Karl Marx. Author H.G. Wells was a member of the Fabian Society from 1903 to 1908. Wells left the Fabian Society because of a controversy that highlighted the direction of the group in the early 1900s.

Pease reported (Pease, 1916, Chapter IX) that Wells presented a paper titled "Faults of the Fabians" to the Fabian Society in February 1906. The presentation marked the beginning of a dispute between Wells and other members of the Fabian Society. Wells supported a more activist agenda, arguing that the Fabian Society was not asserting its agenda enough to reshape society. Within the year, Fabian Society members were motivated to reevaluate the British political system by Scottish socialist Keir Hardie and the Independent Labour Party.

Hardie was the first socialist elected to the British House of Commons in 1892. He contributed to the formation of the Independent Labour Party, or ILP, in 1893. As a result of his activities, Hardie lost the support of the Liberal Party and his parliamentary seat.

Three socialist organizations joined forces with labor unions in 1900 to form the Labour Representation Committee (LRC) in Britain. The LRC consisted of two members from the Independent Labour Party (ILP), two members of the Social Democratic Federation (SDF), one member from the Fabian Society, and seven members of trade unions. Hardie was a member of the LRC from the ILP. Pease was a member of the LRC from the Fabian Society, and Ramsay MacDonald, a member of the Fabian Society and the ILP, served as secretary of the LRC.

Hardie was one of two ILP members who were elected to the House of Commons in 1900. The ILP won 29 seats in the House of Commons in 1906. Two of the seats were occupied by Hardie and MacDonald. That same year, the group changed its name from the LRC to the Labour Party. Hardie was elected Chair of the Labour Party, and MacDonald served as Secretary of the Labour Party (Simkin, 2017). In 1924, Ramsay MacDonald was elected as the first British Prime Minister from the Labour Party.

The ILP's election victory in 1906 demonstrated to the Fabian Society that Wells' "proposal for an enlarged and invigorated society came at the precise moment" when British voters were willing to support socialist politicians (Pease, 1916, Chapter IX). As of this writing (2023), British Prime Ministers that were members of both the Labour Party (UK) and the Fabian Society include Ramsay MacDonald, Clement Atlee, Harold Wilson, James Callaghan, Tony Blair, and Gordon Brown. As indicated in Figure 4.1, the Fabian Society is historically linked to national government.

Wells' 1906 proposal and an alternative proposal written by Shaw were submitted to the Executive Committee of the Fabian Society. Both proposals were submitted for consideration to the membership. According to Pease, "the real issue was a personal one…Was the Society to be controlled by those who had made it or was it to be handed over to Mr. Wells? We knew by this time that he was a masterful person, very fond of his own way, very uncertain what that way was, and quite unaware whither it necessarily led. In any position except that of leader, Mr. Wells was invaluable, as long as he kept it! As leader, we felt he would be impossible, and if he had won the fight he would have justly claimed a mandate to manage the Society on the lines he had laid down. As Bernard Shaw led for the Executive, the controversy was really narrowed into Wells versus Shaw" (Pease, 1916, Chapter IX).

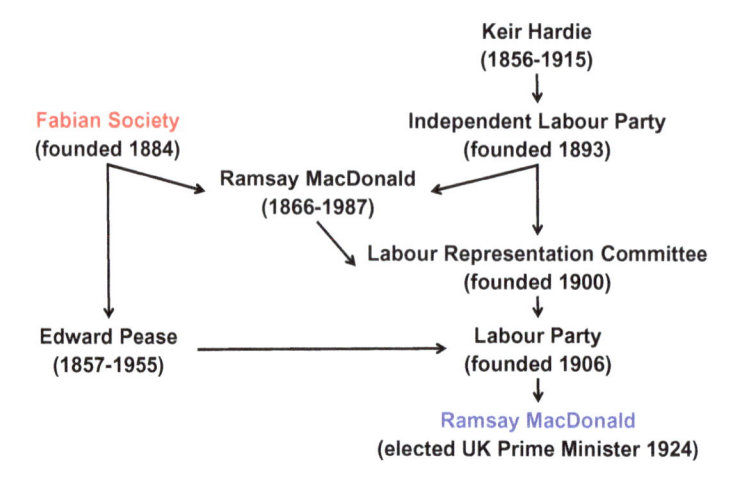

```
                              Keir Hardie
                              (1856-1915)
                                   ↓
    Fabian Society          Independent Labour Party
    (founded 1884)              (founded 1893)
           ↘     Ramsay MacDonald    ↙
                 (1866-1987)
                         ↘
                   Labour Representation Committee
                         (founded 1900)
                              ↓
    Edward Pease   →      Labour Party
    (1857-1955)           (founded 1906)
                              ↓
                        Ramsay MacDonald
                  (elected UK Prime Minister 1924)
```

Figure 4.1 Connections from the Fabian Society to National Governance (Fanchi, 2019)

Wells served on the Executive Committee until he resigned in 1908. Pease wrote that "Mr. Wells was the spur which goaded us on, and though at the time we were often forced to resent his want of tact, his difficult public manners, and his constant shiftings of policy, we recognized then, and we remember still, how much of permanent value he achieved" (Pease, 1916, Chapter IX). Pease called Wells' book *New Worlds for Old* (Wells, 1909) "perhaps the best recent book on English Socialism" (Pease, 1916, Chapter IX). The Fabian Society published *New Worlds for Old* in 1908 while Wells was still a member. The MacMillan Company in New York published it commercially in 1909.

4.2 Fabian Globalism

Leonard S. Woolf, husband of author Virginia Woolf and member of the Fabian Society, published two reports prepared for the Fabian Research Department during World War I (Woolf, 1916). The purpose of the reports was to present a case in support of global governance by a socialist government. Woolfs reports were published with a project by a Fabian Committee for a Supranational Authority that will Prevent War. The reports were published with an introduction by Fabian Society member Bernard Shaw.

Fabian Socialism 55

```
Thomas Davidson  →  Vita Nuova
  (1840-1900)         (founded 1883)
                           ↓
Edward Pease    →    Fabian Society
  (1857-1955)        (founded 1884)
                           ↓
                     Leonard Woolf
                      (1880-1969)
                           ↓
                  Supranational Authority
                     (published 1916)
```

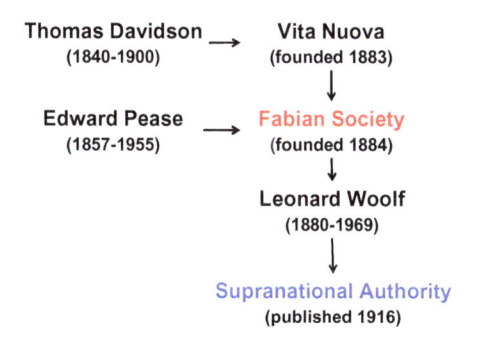

Figure 4.2 Links from the Fabian Society to Globalism (Fanchi, 2019)

In an introduction to Woolf's book, Shaw wrote that "Unless and until Europe is provided with a new organ for supranational action, provided with an effective police force, all talk of making an end of war is mere waste of breath" (Woolf, 1916, p. xv). Furthermore, Woolf wrote that "The alternative to war is law. What we have to do is to find some way of deciding differences between States, and of securing the same acquiescence in the decision as is now shown by individual citizens in a legal judgment. This involves the establishment of a Supranational Authority, which is the essence of our proposals" (Woolf, 1916, p. 372). Links from the Fabian Society to a globalist conception of a supranational authority are depicted in Figure 4.2.

Woolf presented a compilation of articles that supported the establishment of a supranational authority in 1916 (Woolf, 1916, Section II). The formation of an International High Court was the first step needed to develop a supranational organization. The first supranational organization of the 20th century, the League of Nations, was established after World War I. The League of Nations then created the Permanent Court of International Justice in The Hague, Netherlands. The court existed from 1922 to 1946 and was known as the World Court.

Woodrow Wilson, President of the United States, is widely recognized for conceptualizing The League of Nations. Wilson was a key advocate of the League of Nations. The League of Nations failed in the 1930s, in part because the United States Senate chose not to ratify membership. Many Americans opposed the participation of the United States in World War I and did not want to be involved in European politics after the war ended. The failure of the League of Nations did not discourage the creation of another supranational authority after World War II.

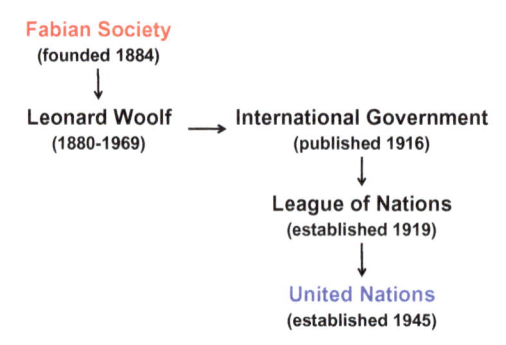

56 *The Energy Transition and the Struggle for Global Governance*

```
        Fabian Society
        (founded 1884)
              ↓
        Leonard Woolf  →  International Government
        (1880-1969)        (published 1916)
                                 ↓
                           League of Nations
                           (established 1919)
                                 ↓
                            United Nations
                           (established 1945)
```

Figure 4.3 Connections from the Fabian Society to the United Nations (Fanchi, 2019)

The Council on Foreign Relations, a subsidiary of the British Round Table (Samuelson, 2017), and the Fabian Society helped found the United Nations in 1945 following World War II. The United Nations Charter authorized the World Court to become the International Court of Justice in 1945. Figure 4.3 depicts the links between the Fabian Society and the United Nations.

Chapter 5
Road to Globalism

In his book *The Battle*, Arthur C. Brooks highlighted two competing and irreconcilable visions of globalism in the United States (Brooks, 2010, p. 1). He contended that the American people are engaged in a political war between big government and free enterprise. Free enterprise, according to Brooks, is "the system of values and laws that respects private property, encourages industry, celebrates liberty, limits government, and creates individual opportunity" (Brooks, 2010, p. 3). On the other hand, Brooks viewed big government as the democratic socialism favored by Fabian Socialists. Democratic socialism in Europe is being implemented by large government agencies exercising control over the commercial sector. In addition, we saw in the previous chapter that Fabian Socialists want to establish a supranational authority as the global government. Further insight into the history of global governance is presented here.

5.1 World War I and the Inquiry

Woodrow Wilson's 14-point plan to end World War I included a recommendation to establish the first supranational authority of the modern age, the League of Nations. In this section, we discuss the intended role of the League of Nations and why the United States chose not to become a member.

Wilson defeated then-President William Howard Taft as well as previous President Theodore (Teddy) Roosevelt in a three-way race for President of the United States in 1912. Teddy Roosevelt was President from 1901 to 1909, Taft from 1909 to 1913, and Wilson from 1913 to 1921. World War I began in 1914 during Wilson's first term.

In May 1915 a German submarine sank the British passenger ship Lusitania. British and American passengers on the Lusitania were killed in the attack. Germany announced that they would attack any shipping in the waters surrounding the British Isles by designating the area a war zone. Germany was aware that the United States favored Britain. Furthermore, Wilson realized that the American people were not ready to participate in what many saw as a European war. When Wilson sought a second term in 1916, he touted legislative successes such as limiting the length of a workday for railroad workers to 8 hours and campaigned to keep the United States out of World War I by adopting the anti-war slogan "he kept us out of war."

Following election to his second term, Wilson used the German doctrine of unlimited submarine warfare to justify entering the war in support of Britain, France, and the other Allies. Wilson asked Congress to declare war on Germany on April 2, 1917. By December 1917, it was evident that the Allies would prevail. Wilson convened a committee of specialists under the chairmanship of Wilson's advisor Edward M. House (1858–1938). House's committee was known as The Inquiry and was charged with drafting recommendations for a comprehensive peace agreement.

5.2 Walter Lippmann's Brush with Fabian Socialism

In 1906, Walter Lippmann enrolled at Harvard University to study philosophy and languages (German and French). In 1909, he presided over the Harvard University chapter of the Intercollegiate Socialist Society (ISS) (Shafer and Snow, 1962). One objective of the ISS, which was founded in 1905, was to advocate for socialism on higher education campuses. Its founders included Jack London and Upton Sinclair, who would later become acclaimed authors. At the time of its founding, both London and Sinclair were in their twenties. London served as the ISS's inaugural president.

Walter Lippmann completed the requirements for a Bachelor of Arts degree at Harvard in 3 years and chose to remain a fourth year working as assistant to Spanish-American philosopher and poet George Santayana. Lippmann was exposed to Fabian Socialism when he attended a course led by British visiting lecturer and former Fabian Socialist Graham Wallas (1858–1922).

Wallas and his friend Sidney Webb were early members of the Fabian Society. Webb was encouraged to join the Fabian Society by George Bernard Shaw in 1885 (BOE Webb, 2017). Within a few years, Webb published essays about socialism, became a member of the Fabian Society executive committee, and met Beatrix Potter. Webb married Potter in 1892.

A few years after their marriage, in 1895, Sidney and Beatrix Webb established the London School of Economics and Political Science (LSE). The LSE's mission was to educate students in economics with a more socialistic orientation. The Webbs asked Graham Wallas to serve as the LSE's inaugural director. Although Wallas declined the offer, he decided to teach at the LSE where he achieved the rank of professor of politics.

Wallas lost patience with the Fabian Society's seemingly endless debates and came to believe that its founders had developed anti-liberal sentiments. In 1895, Wallas left the executive committee; in 1904, he left the Fabian Society itself.

After earning his Harvard degree in 1910, Walter Lippmann worked for liberal socialist Ralph Albertson as a cub writer at the small left-leaning publication The Boston Common. Lippmann then moved to the national magazine entitled *Everybody's Magazine* where he worked for major byline writer Lincoln Steffens (Whitman, 1974; Duffy, 2009).

When Theodore Roosevelt ran against Taft and Wilson for the presidency in 1912, Steffens and Lippmann supported Roosevelt's Progressive Party. Lippmann then moved from *Everybody's Magazine* to a position as executive secretary for the newly elected socialist mayor George R. Lunn of Schenectady, New York. Lippmann worked a few months for Lunn before focusing on writing.

Together with his friend and fellow journalist Herbert Croly (1869–1930), Lippmann co-founded The New Republic in 1913, the same year he published the book *A Preface to Politics*. Croly, who also received his education at Harvard, rose to prominence as a leader of the progressive movement. The Promise of American Life, written by Croly in 1909, promoted better economic planning, increased investment in education, and the creation of a community founded on "the brotherhood of mankind."

Heiress Theodora Payne Whitney and her husband Williard Straight, a banker and diplomat, invited Croly to help them create The New Republic as a liberal publication that would offer perceptive liberal commentary on politics, international relations, and culture. Straight and Whitney provided funds for the journal. Croly recruited Lippmann to join The New Republic's editorial board with Croly effectively serving as

Editor-in-Chief. In November 1914, The New Republic released its first article (History.com New Republic, 2007).

The New Republic maintained an anti-war position until 1915. In his 1916 book *Drift and Mastery*, Lippmann started to stray from socialism while holding on to progressive principles. Lippmann met Wilson's friend and adviser Edward M. House when he used The New Republic to assist Wilson's reelection.

House persuaded Lippmann, a pacifist, to support Wilson's doctrine of limited military readiness for wartime entry. Lippmann accepted a position as an assistant to Wilson's Secretary of War, Nelson Baker, in 1917. Lippmann was subsequently appointed by the House to serve as secretary of The Inquiry.

The Inquiry consisted of over 100 researchers, executives, and staff assistants. Academics from fields like geography, political science, history, economics, and international law made up a large portion of the researchers. Lippmann emerged as one of the main designers of Wilson's strategy to end World War I. Links between the Fabian Society and Woodrow Wilson's 14 Points are depicted in Figure 5.1.

5.3 Wilson's 14 Points and the League of Nations

President Woodrow Wilson delivered his 14-point plan to end World War I to the United States Congress on January 18, 1918 (Wilson 14, 1918).

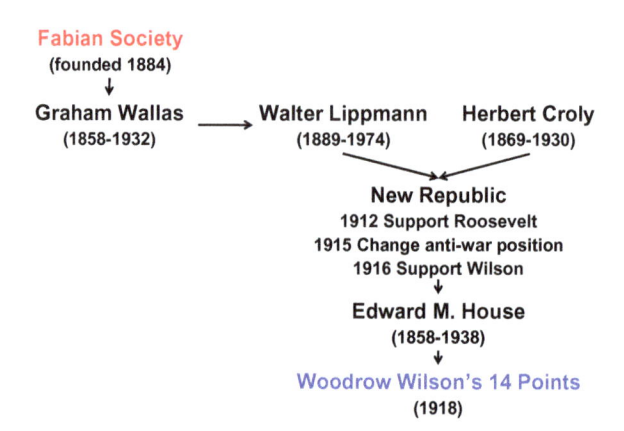

Figure 5.1 Connections from the Fabian Society to Wilson's 14 Points (Fanchi, 2019)

In his preamble, Wilson declared that the "day of conquest and aggrandizement is gone by; so is also the day of secret covenants entered into in the interest of particular governments." He said that the United States "entered this war because violations of right had occurred which touched us to the quick and made the life of our own people impossible unless they were corrected and the world secure once for all against their recurrence." Wilson believed that the world should "be made safe for every peace-loving nation which, like our own, wishes to live its own life, determine its own institutions, be assured of justice and fair dealing by the other peoples of the world as against force and selfish aggression. All the peoples of the world are in effect partners in this interest…" Wilson then introduced his 14-point plan as the only viable program that could assure world peace.

International agreements, freedom of navigation of the seas, international trade, national armaments, and the impartial adjustment of colonial claims were the subject of points 1 through 5. Territorial claims in such countries as Russia, Belgium, France, Italy, Austria-Hungary, Rumania, Serbia, Montenegro, the Balkan States, Turkey, and Poland were addressed in points 6 through 13. Point 14, the last point, proposed a supranational authority: "A general association of nations must be formed under specific covenants for the purpose of affording mutual guarantees of political independence and territorial integrity to great and small states alike."

In October 1918, the German government agreed to a general armistice in accordance with Wilson's 14-point plan. In November 1918, Germany signed the Armistice.

Wilson attended the 1919 Paris Peace Conference. Many aspects of Wilson's 14-point plan were rejected by the Allies. Instead, they wanted the peace treaty to include reparations for war damages paid by Germany and limits on the territorial size of Germany.

The Treaty of Versailles was signed on June 28, 1919, at the Palace of Versailles. It established the League of Nations Covenant, the Permanent Court of International Justice, and the International Labor Organization.

Walter Lippmann, a United States delegate at the Paris Peace Conference, did not agree with the amendments made to Wilson's 14-point plan. Lippmann and his New Republic colleague Herbert Croly did not support the peace treaty that was adopted at the Paris Peace Conference. They began to distance the New Republic from Wilson by encouraging opposition to the Treaty of Versailles and rejection of United States participation in the League of Nations.

Lippmann believed that the League of Nations would be significantly influenced by socialist members. He wrote in 1919 that "It will be difficult enough in all conscience to secure harmony in a League when half the world is socialist and the other half anti-socialist" (Lippmann, 1919, p. 63). Lippmann was concerned that the lack of harmony in the League of Nations would render the League irrelevant (Lippmann, 1919, p. 64): "…if the League is not to find itself marooned on the dry sands of irrelevance it should take steps to introduce into its own structure the conciliatory influence of the opposition parties."

On January 10, 1920, the Treaty of Versailles came into force. It contained Allied demands that were not part of Wilson's 14-point plan. The United States Senate had to ratify the treaty but chose not to do so in March 1920. As a result, the United States did not become a member of the League of Nations.

Adolph Hitler and the National Socialists (Nazis) gained power in Germany in the 1930s in part because of the punitive provisions of the Treaty of Versailles. Two provisions were particularly concerning: demand for German reparations, and territorial limitations. In addition, the League of Nations was unable to garner widespread international support. The rise of the United Nations, the second supranational organization created in the 20th century, would not appear until the conclusion of World War II.

5.4 The Marxist View of Environmentalism Goes Global

Political strategies for achieving social progress in the political world include the establishment of a government with the authority to grant and restrict rights, manage the economy, redistribute wealth, and define morality. The League of Nations was the first effort in the 20th century to create a governing body that could settle international disputes. Woodrow Wilson supported the formation of the League of Nations in Geneva, Switzerland after World War I, but its inability to effectively function as a global governing body and its failure to prevent a second world war led to the establishment of the United Nations in New York City in 1945.

Karl Marx and Vladimir Lenin set the stage for the emergence of Josef Stalin (1878–1953). The line of succession of Soviet prime ministers began with Lenin (1923–1924) followed by Alexei Rykov (1924–1930), Vyacheslav Molotov (1930–1941), and then Stalin (1941–1946). Stalin

```
Georg Hegel
(1770-1831)
    ↓
 Karl Marx
(1818-1883)
    ↓
Vladimir Lenin
(1870-1924)
    ↓
Joseph Stalin
(1878-1953)
    ↓
USSR joins UN (1945)
```

Figure 5.2 Connections from Georg Hegel to the UN (Fanchi, 2019)

was Prime Minister when the Soviet Union joined the United Nations as a charter member.

The United Nations established a Security Council as a means of creating and maintaining international order. The United Nations Security Council included five permanent members: China, France, Great Britain, the Soviet Union, and the United States. These countries were seen as the victors of World War II. Each permanent member of the UN Security Council was granted veto power over any resolutions introduced in the Security Council. When the Soviet Union disbanded in 1991, its position in the United Nations was taken over by the Russian Federation. Other republics of the former Soviet Union were allowed to seek membership in the United Nations as sovereign states.

One intellectual path that led to the founding of the United Nations began with Georg Hegel. Hegel believed that worldviews founded on religious convictions and faith should be replaced by a new worldview based on scientific reason and truth. We pointed out in Section 3.2 that Karl Marx was influenced by Hegel's views. Figure 5.2 summarizes connections from Hegel to the United Nations. As we shall see, proponents of a one-world government have used the United Nations as a forum to advance their globalization agenda.

Chapter 6

U Thant Weaponizes the Environment

The United States and the developed nations of Europe led the United Nations from its inception after World War II until U Thant assumed the post of acting secretary-general in November 1961. Prior to U Thant's selection as secretary general, the primary concern of the United Nations was the Cold War between the Soviet Union and the United States. U Thant believed he could use the environment to implement a UN agenda to advance social progress.

U Thant came from a developing country. On January 22, 1909, he was born in the small Burmese village of Pantanaw, where he later became an educator. U Thant was selected as Burma's representative to the UN in 1957. His position had the rank of ambassador. He held the position until he became interim secretary general in 1961. The selection of U Thant as UN secretary general heralded the emergence of the third world as a political force within the United Nations.

U Thant believed that third-world countries could act as a group of non-aligned countries. As a group, the non-aligned bloc had the potential of rivaling the two Cold War blocs led by the United States and the Soviet Union. The UN under U Thant's administration started to focus more on the north-south split between rich and developing countries and less on the Cold War. As secretary general, U Thant encouraged more developed countries to increase their contributions to the UN budget to fund programs supported by member states. In addition, he asked more developed countries to share their wealth with less developed countries, and he urged countries with colonies to give up control of colonial lands.

The constraints of his position were a source of dissatisfaction for U Thant. For instance, he could refer to UN resolutions that opposed colonialism but were unable to compel the expulsion of a colonial power; he could advocate for the redistribution of wealth globally but could not impose taxes on member states; and he could support the implementation of peacekeeping operations but could not make member states provide enough funding to maintain the peace. He recognized, however, that public interest in environmental issues could help him acquire more global influence.

The public showed growing concern about the impact of human activity on the environment in the 1960s. One of the first people to warn the public about the possible detrimental impact of human activity on the environment was Rachel Carson. She publicized the adverse effects of chemical pesticides in her book *Silent Spring* (Carson, 1962). In 1968, public concern about the environment motivated the General Assembly to recommend that the UN start gathering information on the state of the environment and propose preventive actions (Mische and Ribeiro, 1998).

U Thant realized that UN member states were not unified by peacekeeping and economic efforts. It was his hope that member states would unite to counter the threat of worldwide pollution. In May 1969, he told the UN General Assembly that environmental disaster would occur within 10 years unless member states took steps to avert disaster. A month later he blamed most of the problems on the United States (Lewis, 1985). The Environmental Protection Agency was established by United States President Richard Nixon in 1970 in response to public concern about the environment.

In the opening of his 1969 study *Man and His Environment* (UN, 1969), U Thant emphasized the worldwide scope of the threat. In a 1970 article (U Thant, 1970), he discussed environmental issues and advocated for a new global order. According to U Thant, humanity faced "not merely a threat, but an actual worldwide crisis involving all living creatures, all vegetable life, the entire system in which we live, and all nations large or small, advanced or developing." Furthermore, he said that humanity now faces "a rapidly increasing imbalance between the life-sustaining systems of the Earth and the demands, industrial, agricultural, technological, and demographic, which its inhabitants put upon it" (U Thant, 1970, p. 13).

U Thant cited urbanization and increasing population as the two main factors contributing to the environmental issues. He said that the "unthinking exploitation and abuse of the world's natural resources, and the plunder, befouling, and destruction of our native Earth, have already gone too far for us to rely any more on pious hopes, belated promises, and tardy efforts at self-discipline" (U Thant, 1970, p. 16).

U Thant argued that a global authority was needed to enable prompt and effective action to solve environmental problems. The global authority should be closely associated with the UN and be able to implement and enforce decisions.

According to U Thant, "nothing less than a new step toward world order would do" (U Thant, 1970, p. 17) to save the environment. He called for globalism by asking "Do the sovereign nations of the world have the courage and the vision to set up and support such an agency now, and thus, in the interest of future generations of life on Earth, depart radically from the hitherto sacred paths of national sovereignty?" (U Thant, 1970, p. 16) Furthermore, he asked, "Is it unrealistic to suggest that the undoubted global challenge we now face might become the basis for a new start in world order and a more civilized and generous way of life for the peoples of the Earth?" (U Thant, 1970, p. 17).

A suggestion to organize an international environmental conference was proposed by Sweden in 1967 as a strategy for advancing social progress. U Thant endorsed the request for an international environmental conference in 1969. He used the United Nations Educational, Scientific, and Cultural Organization (UNESCO) to advance his environmental agenda. UNESCO held regional symposia on the environment in 1969 and 1971 (Mische and Ribeiro, 1998). In 1972, UNESCO held a world conference in Stockholm, Sweden that became known as the 1972 Stockholm Conference. The conference was chaired by Maurice Strong (1929–2015), who was emerging as a key player in connecting environmentalism to global government. His story is presented in Part 3.

Part 3
Weaponizing Climate Change

Chapter 7

What Is Anthropogenic Climate Change?

The material in Part 2 showed how the environment could be weaponized to help establish a global government. In this part, Part 3, we describe how the weaponization of climate change can be used to empower a global government. We summarize the arguments in support of anthropogenic climate change in Chapter 7 and review the debate over the role of humans in climate change in Chapter 8. Our account of the emergence of global environmental socialism continues with a discussion of Maurice Strong's influence on the United Nations in Chapter 9. We show in Chapter 10 that Barack Obama continued the movement toward global environmental socialism when he became President of the United States in 2009.

7.1 Anthropogenic Climate Change Defined

Climate change seems to be the result of two major factors: natural causes, and human activity. The relative impact of these factors on climate change is still being determined. People who believe climate change is due primarily to human activity are known as proponents of anthropogenic, or human-caused, climate change. They bolster their views using evidence from such factors as the release of greenhouse gases, variations in sea level, and variations in Arctic Sea ice. The latter two factors are connected because an increase in global temperature is likely to cause the melting of Arctic Sea ice and increase sea levels worldwide. Lindsey and

Dahlman (2022) predict that the average global temperature will rise approximately 0.36°F (0.22°C) per decade.

Many international attempts have been made to mitigate the impact of human activity on the climate. In the following section, we review methods designed to mitigate greenhouse gas emissions.

7.2 Mitigating Greenhouse Gas Emissions

Atmospheric carbon dioxide concentration has been measured since 1958 using the Keeling curve. Charles David Keeling began measuring atmospheric carbon dioxide concentration at the Mauna Loa Observatory on the Big Island of Hawaii. The crest of the Mauna Loa volcano is higher than sources of pollution at lower elevations. Several astronomical observatories have been built atop Mauna Loa because there is very little light pollution. It is also a good location for measuring atmospheric gas composition.

The Mauna Loa Keeling curve is shown in Figure 7.1. A steady increase in atmospheric carbon dioxide concentration has been observed

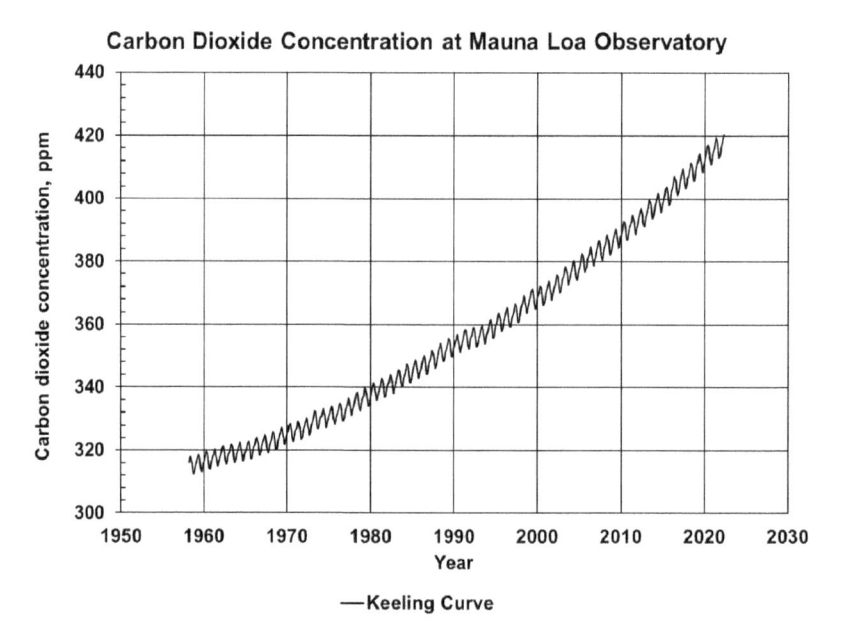

Figure 7.1 The Keeling Curve (Scripps Keeling, 2022)

at Mauna Loa since the first measurements were made in 1958. The saw tooth structure in the curve shows that atmospheric carbon dioxide concentration varies cyclically each year but increases from year to year. Measurements of atmospheric carbon dioxide concentration at Mauna Loa began at approximately 310 parts per million (ppm) and have increased to more than 400 ppm. For comparison, the pre-industrial concentration of atmospheric carbon dioxide was approximately 288 ppm.

A forecast of atmospheric carbon dioxide concentration throughout the 21st century was presented by Wigley *et al.* (1996). They argued that the concentration of atmospheric greenhouse gases should be reduced to avoid adverse environmental effects. According to research by other scientists, plant growth is being facilitated by increasing atmospheric carbon dioxide concentration in a "greening earth" effect (Chen *et al.*, 2020). The ideal level of atmospheric carbon dioxide is unknown.

7.2.1 *Geologic Sequestration*

Geologic sequestration refers to the capture and storage of greenhouse gas emissions in geologic formations. For example, carbon dioxide can be stored in coal beds. The injection of carbon dioxide into coal beds displaces coal bed methane because coal preferentially absorbs carbon dioxide. The displaced methane can be produced as a commercial resource. If the methane is then burned as a fuel, it produces carbon dioxide as a combustion byproduct. The resulting carbon dioxide can then be sequestered.

Another sequestration technique is to inject greenhouse gases, especially carbon dioxide, into hydrocarbon reservoirs. If the gas is injected at sufficient pressure, it can mix with, or be miscible with, oil in the reservoir. Miscible gas injection swells the oil and displaces it towards production wells. Even immiscible gas injection, which is achieved at lower pressures than miscible gas injection, can help displace oil in the reservoir, but not as effectively as miscible gas injection.

If gas is injected into an oil reservoir, oil recovery can be enhanced; but the operator must be careful not to exceed fracture pressure. Reservoir rock can fracture if injection pressure is greater than fracture pressure. Fractures in reservoir rock can form undesired pathways for fluid flow. It is possible that the new pathways can bypass oil in the reservoir or allow fluid to flow out of the reservoir.

A salt dome is another geologic structure that can be used to store greenhouse gases. An impermeable storage tank can be formed by injecting water into the dome to create a cavern. The amount of carbon dioxide that can be stored in the cavern depends on the volume of the cavern.

Geologic sequestration is an example of Carbon Capture and Storage (CCS). When captured carbon can be put to work, the process is an example of Carbon Capture, Utilization, and Storage (CCUS). Injecting carbon dioxide into an oil reservoir to improve the displacement of oil and enhance oil recovery is an example of CCUS. CCS and CCUS technologies are considered "negative emissions" technologies because they remove carbon dioxide from the atmosphere.

Point to Ponder: Is Geologic Sequestration a Desirable Policy?

Geologic sequestration seems to be a desirable policy for a society that wants to minimize carbon dioxide emissions while it continues to use fossil fuels. Opponents of geologic sequestration argue that it is an undesirable policy because it can reduce a society's motivation to develop an energy infrastructure that replaces fossil fuels with sustainable energy sources.

Another factor to be considered is the cost geologic sequestration adds to fossil fuel electricity generation. The additional cost of capturing and storing carbon dioxide emitted by the consumption of fossil fuels can help make the cost of using renewable energy sources more competitive.

7.2.2 Alternatives to Geologic Sequestration

A few alternatives to storing greenhouse gases in geologic structures are considered here.

Ocean hydrate storage refers to the storage of greenhouse gases, especially carbon dioxide and methane, in ocean hydrates. Greenhouse gases can be positioned in regions of the ocean where the combination of low-temperature and high-pressure conditions leads to the formation of a hydrate from a mixture of water and gas. Hydrates are structures that bond a molecule such as carbon dioxide within a ball-like enclosure formed by water molecules. A concern with ocean hydrate storage is that ocean warming could lead to the breakdown of the hydrate structure and the release of the bound gas molecule. As a result, the greenhouse gas contained in the hydrate would be released into the atmosphere and contribute to the greenhouse effect.

Captured carbon dioxide can be used in processes that typically rely on carbon dioxide. For example, carbon dioxide is used in plastic or concrete materials; it can be fed to algae to produce biofuels such as algal oil; and it can be used in an Integrated Electricity-Natural Gas System (IEGS). The IEGS generates electricity using renewable technology such as a wind turbine. Some of that electricity is then used to synthesize methane from hydrogen and carbon dioxide in a Power to Gas (P2G) facility. From there, the synthesized methane is subsequently used in gas turbines to generate additional electricity (Hu *et al*., 2022).

Surface water (H_2O) in lakes and oceans can react with atmospheric carbon dioxide (CO_2) to create carbonic acid (H_2CO_3). The amount of carbon dioxide in the atmosphere can increase the amount of carbon dioxide dissolved in water. The carbonic acid formed by the reaction of carbon dioxide and water makes lakes and seas more acidic. An increase in acidity can damage lake and sea life and have a negative impact on the human food supply.

The utilization of chemical or physical processes to extract carbon dioxide directly from the atmosphere is called direct air capture (DAC). DAC technologies are considered sequestration technologies if the extracted carbon dioxide is placed in safe, long-term storage.

7.3 Climate Change and Adaptation

There is disagreement about the extent that climate is being adversely affected by greenhouse gas emissions. Proponents of anthropogenic climate change support the implementation of technologies that can mitigate the effect of greenhouse gas emissions. People who do not believe human activity is the principal cause of climate change may still support the mitigation of greenhouse gas emissions as a precaution. Others contend that it is already too late to stop anthropogenic climate change because it has already started.

English author and environmentalist James Lovelock proposed the earth feedback hypothesis in the 1960s. Lovelock hypothesized that the earth's physical environment works together with its biosphere, including humanity, to provide a life-sustaining environment. This hypothesis has become known as the Gaia hypothesis after Gaia, the ancestral mother of all life in Greek mythology. It is reasonable to believe that the relationship between the earth's physical environment and its biosphere will be affected by global anthropogenic climate change.

In his book *The Vanishing Face of Gaia*, Lovelock raised concerns about the veracity of climate predictions (Lovelock, 2009, p. 68). He believed the most important question we should be asking about climate change was "How much and how fast is the earth heating?"

Lovelock believed that anthropogenic climate change was occurring and claimed that a reliable measure of the earth's thermal balance was the change in sea level. He attributed sea level rise to the expansion of a warming ocean and water flowing into the ocean from ice melting on land. The decreasing size and eventual disappearance of glaciers are indicators of ice melting on land. Lovelock concluded that humans should prepare to adapt to changing sea levels because anthropogenic climate change is inevitable. He said that "Until we know for certain how to cure global heating, our greatest efforts should go into adaptation, to preparing those parts of the earth least likely to be affected by adverse climate change as the safe havens for a civilized humanity" (Lovelock, 2009, p. 68).

> **Point to Ponder: Is Sea Level Change a New Phenomenon?**
>
> Scientists have known that sea level changes have taken place throughout the history of the earth. One measure of sea level is called eustatic sea level. It is the distance from the center of the earth to the surface of a body of water. Several natural factors can contribute to changes in eustatic sea level, such as changes in glaciation, or changes in the rate of continental separation due to the spreading of a mid-ocean ridge. Figure 7.2 shows the author in front of water flowing downhill from a melting glacier near Saas Fe, Switzerland in 2004.

7.4 The Kyoto Protocol

An international treaty known as the Kyoto Protocol was negotiated in Kyoto, Japan in 1997. It set limits on the quantity of greenhouse gases a

Figure 7.2 Water Flowing from a Melting Glacier near Saas Fe, Switzerland (© Fanchi, 2004)

country could emit into the environment. Limits on greenhouse gas emissions set by the treaty were considered by the representatives of some countries to be so restrictive that they could damage their economies without solving the problem of global warming. Furthermore, the treaty was not uniformly applied to every nation. Greenhouse gas emission controls were more restrictive on developed nations than in developing nations. For example, the two most populous countries on Earth — China and India — were rapidly expanding their fossil-fuel-based economies, yet were considered developing, rather than developed, nations. The United States did not sign the Kyoto Protocol because of these concerns.

In December 2009, a conference in Copenhagen attempted to establish another international agreement to manage the emission of greenhouse gases into the atmosphere. A non-binding declaration of intent to limit atmospheric greenhouse gas emissions emerged from the Copenhagen conference.

An agreement to control greenhouse gas emissions was signed by the United States and China in November 2014. Figure 7.3 shows US and

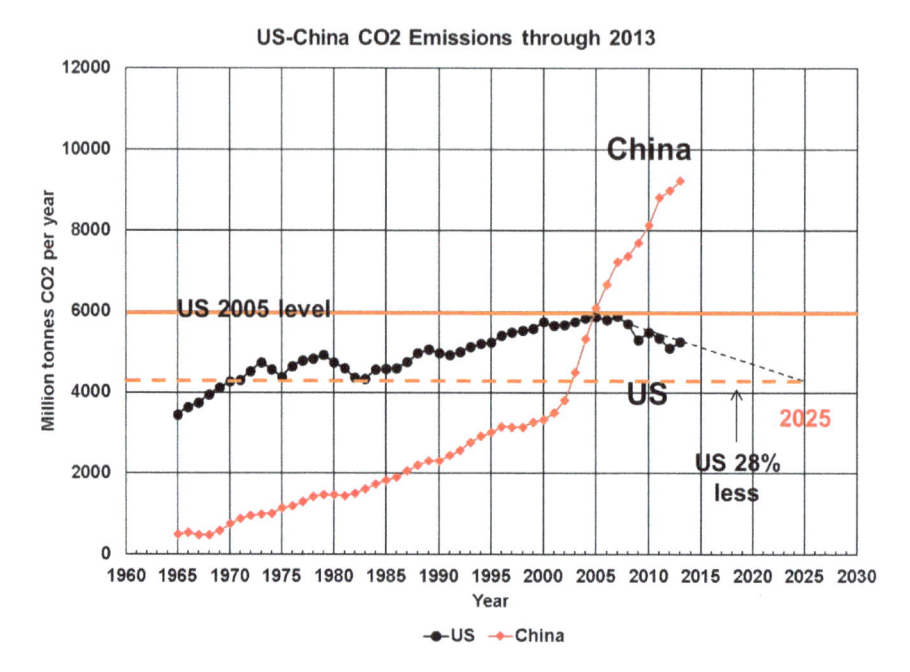

Figure 7.3 US and China CO_2 Emissions Through 2013

China CO_2 emissions prior to the 2014 US–China Agreement. The United States agreed to reduce emissions by up to 28% below its 2005 levels by the year 2025. By contrast, China would not have to reduce its emissions before 2030, but it would have to set the peak of its emissions by 2030. China did agree to increase its share of energy consumption from non-fossil fuels to approximately 20% of total consumption by 2030.

Figure 7.4 shows how well the 2014 US–China Agreement has been working through 2022. US CO_2 emissions have tended to decline since the agreement was signed. China's CO_2 emissions dipped around 2016 and continued to rise since then.

Figure 7.5 shows CO_2 emissions in the global context. The figure shows CO_2 emissions for the world with and without China CO_2 emissions. It also shows the decline in CO_2 emissions during the 2020 COVID-19 pandemic year. The rate of increase in CO_2 emissions is relatively constant when China's CO_2 emissions are not included. China has continued to build coal-fired power plants to help meet its energy needs.

What Is Anthropogenic Climate Change? 79

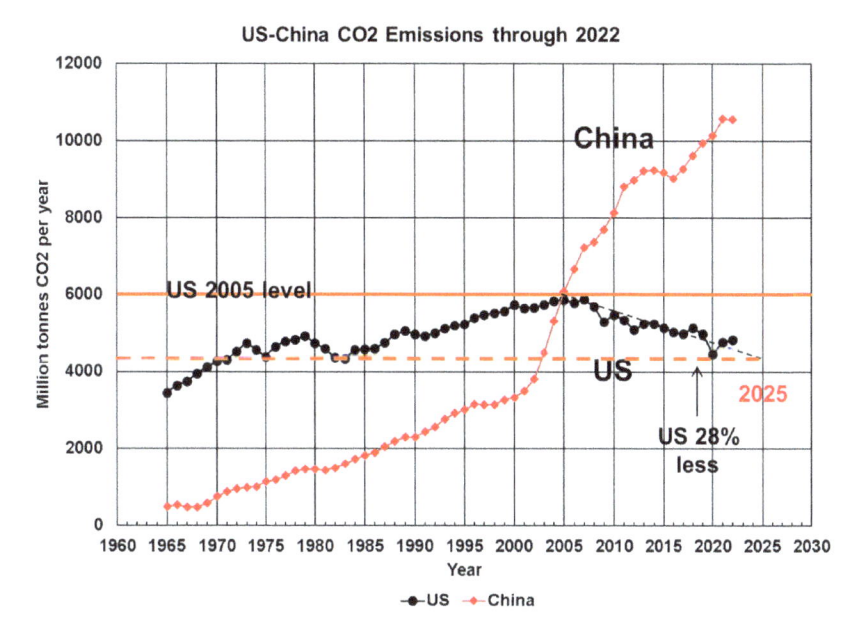

Figure 7.4 US and China CO$_2$ Emissions through 2022

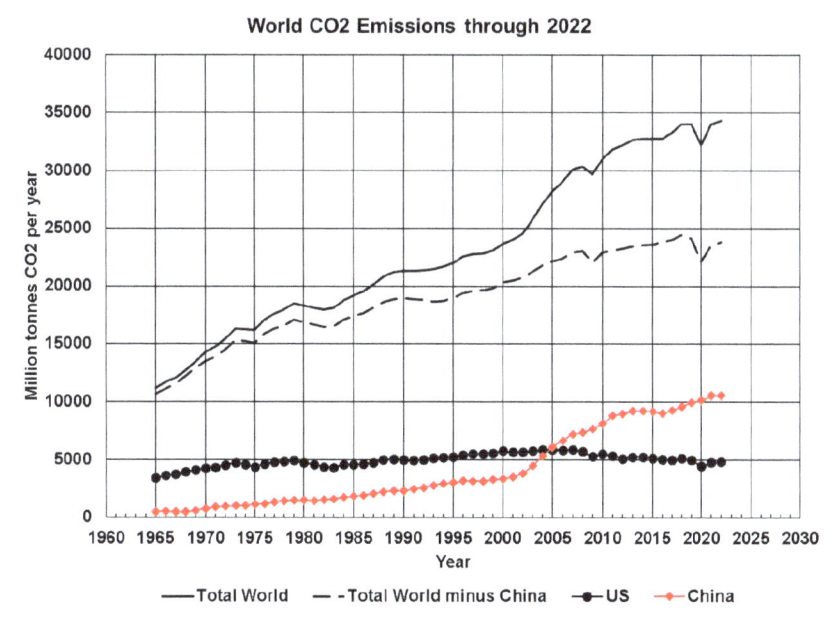

Figure 7.5 World CO$_2$ Emissions Through 2022

7.5 COP21 Agreement — Paris, 12 December 2015

Almost 200 nations met in the city of Paris for a meeting that was held from November 30 to December 11, 2015. The purpose of the meeting was to reach a consensus on a plan to protect the climate from the detrimental effects of human activity. The Paris meeting was known as the 21st Conference of Parties (COP21) to the United Nations Framework Convention on Climate Change (UNFCCC).

One detrimental effect was the emission of carbon dioxide, a greenhouse gas, into the atmosphere. According to the United States Department of Energy (Boden *et al.*, 2017), 70% of global atmospheric carbon dioxide emissions from fossil fuel consumption and some industrial processes were due to the following countries: China (30%), the United States (15%), 28 countries in the European Union (9%), India (7%), the Russian Federation (5%), and Japan (4%). The remaining 30% of global atmospheric carbon dioxide was emitted by all other nations.

Fontinelle and colleagues reported that "an agreement among the leaders of over 180 countries to reduce greenhouse gas emissions and limit the global temperature increase to below 2 degrees Celsius (3.6 Fahrenheit) above preindustrial levels by the year 2100" (Fontinelle, Clarke, and Costagliola, 2022). The agreement is known as the COP21 agreement. It was approved on December 12, 2015 (UN COP21 Agreement, 2015). Key elements of the COP21 agreement are summarized in Table 7.1.

In addition to the elements listed in Table 7.1, up to US$100 billion per year would be provided to developing nations from developed nations. Developing nations were expected to use the funds to reduce emissions

Table 7.1 Key Elements of the COP21 Agreement

A	Governments agree to reduce emissions with a long-term goal of keeping the increase in global average temperature below 2°C above pre-industrial levels.
B	Governments are allowed to determine their own priorities and targets because each country has different circumstances and a different capacity to undertake change.
C	Governments agree to be transparent by providing reports to each other and the public.
D	Governments agree to be adaptable.

and increase their capacity to withstand the effects of anthropogenic climate change. Other nations were urged to voluntarily support less developed nations. The amount of funds provided by developed countries could be changed in the future.

Point to Ponder: Does the COP21 Agreement Require Global Cooperation?

On December 9, 2015, then-United States Secretary of State John Kerry made the following observation:

"The fact is that even if every American citizen biked to work, carpooled to school, used only solar panels to power their homes, if we each planted a dozen trees, if we somehow eliminated all of our domestic greenhouse gas emissions, guess what — that still wouldn't be enough to offset the carbon pollution coming from the rest of the world.

"If all the industrial nations went down to zero emissions — remember what I just said, all the industrial emissions went down to zero emissions — it wouldn't be enough, not when more than 65 percent of the world's carbon pollution comes from the developing world."

As a representative of the Obama-Biden Administration, Kerry made it clear that global cooperation was needed to effectively address anthropogenic climate change.

7.6 American Response to the COP21 Agreement

The 2015 COP21 Paris Climate Agreement was signed by the Obama–Biden Administration in the United States. The COP21 agreement was not a law in the United States because it was not approved by the United States Senate, which is charged with approving international agreements. Failure to secure Senate approval meant that support of the COP21 agreement by the United States could be withdrawn by a new president without an act of Congress.

The Obama-Biden Administration ended in January 2017 when Donald Trump was inaugurated as President of the United States. Trump withdrew from the COP21 Paris Climate Agreement in June 2017. The Trump Administration pointed out that countries such as China and India were being treated as developing nations even though they were responsible for over a third of global atmospheric carbon dioxide emissions. It appeared to the Trump Administration that American workers and taxpayers were being unfairly treated by the COP21 agreement. The Trump Administration believed the cost of the COP21 agreement on the American economy would outweigh its benefits on the global environment.

Joe Biden replaced Donald Trump as President in January 2021. He appointed Antony Blinken to serve as his Secretary of State. Blinken previously served in the Obama–Biden administration as deputy national security advisor (2013–2015) and deputy secretary of state (2015–2017). In a press release dated February 19, 2021, Blinken said that "On January 20, on his first day in office, President Biden signed the instrument to bring the United States back into the Paris Agreement. Per the terms of the Agreement, the United States officially becomes a Party again today.

"The Paris Agreement is an unprecedented framework for global action. We know because we helped design it and make it a reality. Its purpose is both simple and expansive: to help us all avoid catastrophic planetary warming and to build resilience around the world to the impacts from climate change we already see" (Blinken, 2021).

The American response to the COP21 Paris Climate Agreement demonstrates that competing views of anthropogenic climate change and shifts in political power within a nation can impact international agreements.

Chapter 8

Is the Climate Change Debate Settled?

Advocates of anthropogenic climate change (ACC) want society to transition to a low-carbon economy as soon as possible (as an example, see Srivastava and Ramamurthy, 2021). If atmospheric greenhouse gas emissions continue to increase, say advocates of ACC, then the sea level could be up to 23 inches higher in 2100 than it was in 1990. According to the National Oceanographic and Atmospheric Administration (NOAA) in the United States, sea level has been rising at a rate of approximately 12 inches per century, or about one-eighth of an inch per year (NOAA Sea Level, 2023).

Sea level rise can be put into perspective by recognizing that hurricane storm surge can be 20 feet higher than sea level (NOAA Storm Surge, 2023). Flooding from storm surges normally subsides after a few days. By contrast, sea level rise due to climate change can have lasting consequences and may require people to adapt to a long-term change in sea level.

The cost of transitioning every country to a low-carbon economy from a carbon-based economy is immense. The counterargument is that the cost of failing to transition will also be immense. Should nations begin an immediate transition to low-carbon economies, or do we have time to make the transition? The Goldilocks Policy for energy transition used historical data to suggest that nations should be able to make a relatively smooth transition in approximately 60 years. This chapter reviews the debate between an immediate transition and a gradual transition.

8.1 UN IPCC and NIPCC

The United Nations Intergovernmental Panel on Climate Change (UN IPCC) was created by the United Nations Environment Programme (UNEP) and the World Meteorological Organization (WMO) in 1988. That same year, the UN General Assembly endorsed the establishment of the IPCC by WMO and UNEP. The UN IPCC website says that the IPCC "was created to provide policymakers with regular scientific assessments on climate change, its implications and potential future risks, as well as to put forward adaptation and mitigation options.

Through its assessments, the IPCC determines the state of knowledge on climate change. It identifies where there is agreement in the scientific community on topics related to climate change, and where further research is needed. The reports are drafted and reviewed in several stages, thus guaranteeing objectivity and transparency. The IPCC does not conduct its own research. IPCC reports are neutral, and policy-relevant but not policy-prescriptive. The assessment reports are a key input into the international negotiations to tackle climate change" (UN IPCC Mission, 2023).

The mainstream perception of climate is based on UN IPCC assessments. A second group has formed to independently evaluate UN IPCC assessments of changes in atmospheric gas composition on the biosphere. The second, independent group is named the Non-governmental International Panel on Climate Change (NIPCC). Both the UN IPCC and NIPCC include professionals with reputable credentials.

The UN IPCC and NIPCC agree that the climate is changing but do not agree that human activity is the principal cause of the changes. In the past, the UN IPCC has stated that human activity affects the temperature of the atmosphere, accelerates glacier melting, causes the rise of sea levels, and increases the acidity of ocean water. According to the UN IPCC, these changes are largely the result of fossil fuel combustion (UN IPCC AR4, 2007). In contrast, the NIPCC believes that the impact of human activity on either the environment or human well-being has been exaggerated by the UN IPCC (NIPCC CCR-II, 2013).

According to the NIPCC, global climate models do not correctly represent all of the mechanisms that affect climate. Global climate models are computer programs that are used to analyze climate observations. The validity of these computer programs can be verified by using them to replicate past climate behavior. These tests demonstrate that global climate models do not adequately replicate past climate behavior. Therefore, it is reasonable to conclude that the validity of global climate models is limited and raises concerns about the quality of predictions made by the models (Hourdin et al., 2017; Koonin, 2021).

The UN IPCC and NIPCC differ in their assessment of global climate models. According to the UN IPCC, the models are reliable. By contrast, the NIPCC says the models are not reliable because they do not adequately model all factors that influence climate behavior. The exchange of gas molecules between the atmosphere and the ocean is an example of a factor that is still being studied. Researchers can find model limitations when they attempt to replicate historical data.

Advocates of ACC argue that research is needed to understand the details of climate change. A couple of factors that need to be considered are atmospheric temperature and changes in greenhouse gas concentration. It is important to point out that governments continue to provide a sizable amount of money to study climate change factors. This implies that a number of problems are unresolved. The UN IPCC and NIPCC represent two groups of people who considered the available information and reached different conclusions.

The response of the United States to the COP21 Paris Climate Agreement demonstrates that many individuals still have doubts about the legitimacy of the climate issue. Democrats led by Obama and Biden believed that an urgent transition to renewable energy was needed. Republican Trump believed that the transition to sustainable energy should be a goal, but there was time to proceed. The Goldilocks Policy for an energy transition proposed a transition period based on historical data (Fanchi, 2019). Other suggestions for progressing in an uncertain world are considered below.

8.2 The Oil and Gas Climate Initiative

The Chief Executive Officers (CEOs) of ten multinational oil and gas companies founded the Oil and Gas Climate Initiative (OGCI) in 2014 as a means of addressing ACC (OGCI Founding, 2022). The mission of the CEO-led OGCI is to lead the industry in mitigating climate change. According to OGCI (OGCI Founding, 2023), the goal of OGCI is "to create a US$1 billion-plus fund that invests in companies, technologies, and projects that accelerate decarbonization within energy, industry, built environments and transportation."

The strategy of OGCI is to "work individually and collectively to accelerate action towards a net zero emissions future." According to the OGCI strategy document: "OGCI supports the goals of the Paris Agreement, limiting global warming to well below 2°C and pursuing efforts to limit it to 1.5°C, and recognizes that there is a real urgency to act. We support the need for the world to move to a net zero carbon emission future, also called carbon neutrality" (OGCI Strategy, 2023).

The OGCI founding companies were bp (Britain), CNPC (China National Petroleum Corporation), Eni (Italy), Equinor (formerly Statoil of Norway), Pemex (Mexico), Petrobras (Brazil), Repsol (Spain), Saudi Aramco (Saudi Arabia), Shell (The Netherlands), and Total (France). None of the corporate headquarters of each founding company was in the United States. In September 2018, three American-based oil companies — Chevron, ExxonMobil, and Occidental Petroleum — joined ten original members. Today there are twelve OGCI members: Aramco, bp, Chevron, CNPC, Eni, Equinor, ExxonMobil, Occidental, Petrobras, Repsol, Shell, and TotalEnergies (OGCI Founding, 2023).

8.3 Bjorn Lomborg

Environmentalist Bjorn Lomborg began contributing to the worldwide debate on climate change policy with the publication of *The Skeptical Environmentalist* (Lomborg, 2001). He reviewed more recent data and updated his arguments in his book *False Alarm* (Lomborg, 2021).

Some proponents of ACC have claimed that we must mitigate climate change by 2030 or face climate calamity. Lomborg rejected the urgency of this claim. He said that "We are not on the brink of imminent extinction…in almost every way we can measure, life on earth is better now than it was at any time in history" (Lomborg, 2021, p. 8). He cited increased life expectancy, access to health care, a decline in poverty, and better air and water quality as evidence for his position.

According to Lomborg, "Climate change is real. It is caused predominately by carbon emissions from humans burning fossil fuels…we need to stop exaggerating, stop arguing that it is now or never, and stop thinking climate is the only thing that matters" (Lomborg, 2021, p. 7).

Lomborg pointed out that President Joe Biden "relies heavily on the rhetoric of climate apocalypse" (Lomborg, 2021, p. 226). In a White House Fact Sheet, Biden referred to climate change as an "existential threat" (Biden, 2021). Lomborg said that "this rhetoric is vastly exaggerated" (Lomborg, 2021, p. 226).

Table 8.1 Lomborg's Proposed Solutions to Climate Change

1. Implement a carbon tax, a tax levied on the use of fossil fuels, to discourage the production of greenhouse gas emissions.
2. Invest in green innovation, such as nuclear fusion, nuclear fission, hydrogen production by splitting water molecules, and using algae to convert sunlight and carbon dioxide into algal oil.
3. Adapt to environmental changes.
4. Research geoengineering, that is, technology capable of manipulating the earth's environment. Examples include carbon sequestration or increasing the reflection of solar radiation back into space to cool the atmosphere. Lomborg said that the implementation of geoengineering should be done cautiously so there is time to investigate side effects.
5. Elevate other issues that contribute to human prosperity.

"By focusing most of our attention on climate change," Lomborg said, "we're ignoring other, bigger issues…" that could improve human life (Lomborg, 2021, p. 16). In fact, Lomborg recognized that climate change policy could be detrimental to the quality of human life (Lomborg, 2021, p. 12): "For poor countries, climate policy threatens to crowd out the much more important issues of health, education, jobs, and nutrition." He explained that "In poor countries, higher energy costs harm efforts to increase posterity … Countries in the developing world need cheap and reliable energy, for now mostly fossil energy" (Lomborg, 2021, p. 12).

Lomborg said that a cost-benefit analysis is needed to identify cost-effective solutions to mitigate climate change. Table 8.1 summarizes five possible solutions (Fanchi, 2024). The most promising strategy is to develop green energy technologies that are more affordable than fossil fuel combustion. This strategy would provide economic incentives to convert from fossil fuel combustion to clean energy sources.

8.4 Science and Consensus

Advocates for ACC claimed that "97% of scientists believe that ACC is occurring." This implies that the vast majority of scientists believe that climate change is caused by human activity. Is this a valid interpretation? Here we consider the origin of the claim and its validity.

8.4.1 *The 97% Claim*

The United States National Aeronautics and Space Administration (NASA) contributed to the climate change debate. The NASA Global Climate Change website "presents the state of scientific knowledge about climate change while highlighting the role NASA plays in better understanding our home planet" (NASA Climate Change, 2023). NASA personnel have considered peer-reviewed studies from around the world that illustrate "the accuracy and consensus of research results (in this case, the scientific consensus on climate change) consistent with NASA's scientific research portfolio." Here we consider one paper (Cook *et al.*, 2013) that can be considered the origin of the claim that "97% of scientists believe that ACC is occurring."

Cook *et al.* (2013) investigated 11,944 climate abstracts from 1991 to 2011. To be a climate abstract, the abstract had to include topics "global climate change" or "global warming." The authors observed that 66.4% of abstracts did not express a position on anthropogenic global warming (AGW), 32.6% supported AGW, 0.7% rejected AGW, and 0.3% were unsure. The authors then analyzed the subset of abstracts that expressed a position on AGW. They found that 97.1% of this subset of abstracts "endorsed the consensus position that humans are causing global warming. In the second phase of this study, we invited authors to rate their own papers. Compared to abstract ratings, a smaller percentage of self-rated papers expressed no position on AGW (35.5%). Among self-rated papers expressing a position on AGW, 97.2% endorsed the consensus."

Other interpretations of the data presented by Cook *et al.* (2013) are possible. For example, it is reasonable to say that 66.4% of articles with abstracts matching the topics "global climate change" or "global warming" did not explicitly support AGW. The study by Cook *et al.* (2013) refers to a subset, or smaller group, of scientists willing to publish a position on ACC. Even if the 97% consensus claim does represent the views of scientists who conduct climate research, should a consensus of these scientists mean we should accept their conclusions?

8.4.2 *Does Scientific Consensus Matter?*

According to the NASA Global Climate Change website, "Scientific evidence continues to show that human activities (primarily the human burning of fossil fuels) have warmed Earth's surface and its ocean basins, which in turn have continued to impact Earth's climate. This is based on over a century of scientific evidence forming the structural backbone of today's civilization" (NASA Climate Change, 2023). The website writers recognize that the term consensus refers to "a general agreement of opinion, but the scientific method steers us away from this to an objective framework." They then refer to the scientific method: "In science, facts or observations are explained by a hypothesis (a statement of a possible explanation for some natural phenomenon), which can then be tested and retested until it is refuted (or disproved).

"As scientists gather more observations, they will build off one explanation and add details to complete the picture. Eventually, a group of hypotheses might be integrated and generalized into a scientific theory, a scientifically acceptable general principle or body of principles offered to explain phenomena" (NASA Climate Change, 2023).

We said previously that there is a disagreement between the UN IPCC and the NIPCC. Another organization called the Global Climate Intelligence (CLINTEL) Group has expressed a position on climate change and climate policy. CLINTEL was established in 2019 by science writer Marcel Crok and emeritus geophysics professor Guus Berkhout. It is composed of scientists from around the world, although the CLINTEL website notes that it "is not the number of experts but the quality of arguments that counts" (CLINTEL, 2023). They have prepared a World Climate Declaration based on arguments summarized in Table 8.2.

The arguments in Table 8.2 are used to buttress CLINTEL's declaration that a climate emergency does not exist. They advised European Leaders that "science should strive for a significantly better understanding of the climate system, while politics should focus on minimizing potential climate damage by prioritizing adaptation strategies based on proven and affordable technologies" (CLINTEL Declaration, 2023, p. 1).

Table 8.2 CLINTEL Arguments

1	Natural as well as anthropogenic factors cause warming
2	Warming is far slower than predicted
3	Climate policy relies on inadequate models
4	CO_2 is plant food, the basis of all life on Earth
5	Global warming has not increased natural disasters
6	Climate policy must respect scientific and economic realities

The CLINTEL Group highlighted its concern about climate models with the observation "To believe the outcome of a climate model is to believe what the model makers have put in. This is precisely the problem of today's climate discussion to which climate models are central. Climate science has degenerated into a discussion based on beliefs, not on sound self-critical science. Should not we free ourselves from the naive belief in immature climate models?" (CLINTEL Declaration, 2023, p. 2).

Despite claims of consensus by advocates of ACC, counterarguments are being made by independent organizations and the ACC debate continues. It is important to note here that a consensus of scientists does not mean their views are valid. We can see this in a historical example from physics.

At the beginning of the 20th century, almost every physicist believed in Isaac Newton's concept of absolute time. The acceptance of Newton's concept of absolute time was replaced by a new concept of relative time presented by one physicist, Albert Einstein, in his 1905 paper on special relativity (Fanchi, 2023). The views of the physics community were changed by the work of one physicist. Scientific validity depends on the acquisition and analysis of observations.

Chapter 9

Maurice Strong and Global Environmental Socialism

We saw in Chapter 6 that U Thant began to use the United Nations in the 1960s and 1970s to weaponize the environment on a global scale. He endorsed the 1967 Swedish proposal to organize an international environmental conference as a strategy for advancing social progress. U Thant chose to advance his environmental agenda using the United Nations Educational, Scientific, and Cultural Organization (UNESCO). One of his allies was Maurice Strong. Strong's story is presented in the following sections.

9.1 Strong's Early Career

Maurice Strong told the Canadian magazine Maclean's in 1976 that "I am a socialist in ideology, a capitalist in methodology" (Bailey, 1997). He was born in Oak Lake, Manitoba, Canada at the beginning of the Great Depression in 1929. In 1944, at the age of 14, he left school to work as a deckhand. In 1945, when he was 16 years old, Strong began working as a fur buyer in the Canadian north for Hudson's Bay Company. Strong met the Inuit Eskimos while he was in the north and was introduced to Bill Richardson. Richardson's wife was Mary McColl, a member of the McColl family who helped found McColl-Frontenac, an integrated Canadian oil company.

McColl-Frontenac was created by the merger of McColl Brothers, founded by John McColl in 1873, and Frontenac Oil Refineries. It was incorporated in 1927 as McColl-Frontenac Oil Co Ltd. The company became Texaco Canada Ltd in 1959, and Texaco Canada Inc in 1978 when it merged with Texaco Explorations Canada Ltd.

In 1947, Richardson assisted Strong in establishing important relationships with people like UN Treasurer Noah Monod. Strong met young David Rockefeller (1915–2017) during a short stay at Monod's New York residence. David Rockefeller was the grandson of the founder of Standard Oil, John Davison Rockefeller (1839–1937). Chase Bank, David's employer, had given him the responsibility of managing the UN account. Rockefeller helped Strong secure a job as a low-level security guard at the UN offices in Lake Success, New York. According to Strong, he "had a long and cordial relationship with David in later years" (Strong, 2000, p. 73). A few years later, Strong left his security position and relocated to Winnipeg, Manitoba's capital, where he worked as a securities analyst.

Oilman John (Jack) Gallagher hired Strong to work at Dome Petroleum in 1951. Gallagher had previously worked for Standard Oil of New Jersey. Strong had "the opportunity of learning the business from a more operational point of view" (Manitou-Strong, 2020) while employed at Dome Petroleum. He served in a variety of important roles, including vice president of finance.

Strong changed jobs and relocated abroad in the early 1950s. He investigated possible sites for gas stations in East Africa for Caltex, a petroleum firm that was founded as a partnership between the Texas Company and Standard Oil of California. The Texas Company was later known as Texaco, and Standard Oil of California was renamed Chevron.

9.2 The YMCA and European Connections

Maurice Strong became aware of the YMCA organization in Nairobi, Kenya before leaving East Africa and returning to Canada in 1955. He rejoined Dome and made enough money from stock options (Roberts, 2015) to serve as a volunteer in the YMCA World Service Program. In 1955, he went to Paris to celebrate the YMCA's 100th anniversary.

Strong traveled to Geneva while he was in Europe to meet his distant cousin Tracy Strong. The Manitou Foundation, which was established in 1988 by Maurice Strong's second wife Hanne Marstrand, said that "Strong met Tracy Strong, who was the Secretary General of the World Alliance headquartered in Geneva, Switzerland, and a brother of

Anna Louise Strong, the American journalist whose letters from China had been such a source of Maurice Strong's early interest in China.

"Tracy Strong confirmed that he and (Maurice) Strong did indeed have a family relationship, though somewhat distant. Strong was pleased to meet, too, his son, Robbins, of the World Council of Churches in Geneva" (Manitou-Strong, 2020).

Anna Louise Strong (1885–1970), a relative of Maurice Strong, received her education in the United States and traveled extensively throughout the communist world. Her travels included visits to such countries as the USSR, China, and Eastern Europe. She was a member of the Comintern (Communist International) and was "an enthusiastic supporter of the Russian experiment in communism" (BOE AL Strong, 2023). Her writings included books about her travels in communist countries.

In 1958, Anna Louise Strong made China her home. "She was a close friend of Mao Zedong, whom she had first interviewed in a cave in Yenan province in 1946" (BOE AL Strong, 2023). Her relationship with Maurice Strong helped foster a special bond between Maurice and China.

Maurice Strong was selected to serve on the Canadian YMCA's International Committee in 1958. His affiliation with the YMCA afforded him connections that were beneficial to him in the years to come.

9.3 Strong's Preparation for the United Nations

At the relatively early age of 31, Maurice Strong was leading an oil company. Strong served as chairman of Petro-Canada, chairman of the Canadian Development Investment Corporation, and president of the Canadian Industrial Gas and Power Corporation of Canada. Paul Martin, Senior, Canadian Minister of External Affairs, and Lester Pearson, Prime Minister of Canada, were familiar with Strong's work on corporate boards and in international affairs. Pearson invited Strong to accept a position in the Canadian government as a deputy minister responsible for administering the External Aid program (Manitou-Strong, 2020).

Under Strong's leadership, the External Aid program developed into the Canadian International Development Agency (CIDA). The Manitou Foundation reported that "Strong's work with CIDA gave him new insights into the complexities of development. He was troubled by the environmental and social disruption caused by major infrastructure projects, which CIDA supported. It wasn't long before he became involved with environmental politics" (Manitou-Strong, 2020). The *New York Times* reported that Strong eventually admitted to being "an environmental sinner." He said that "we were running the Earth without a depreciation account, in effect spending our capital" (Roberts, 2015).

Maurice Strong's work at CIDA helped him return to the United Nations as a representative of Canada.

9.4 The United Nations and the 1972 Stockholm Conference

The United Nations Conference on the Human Environment was convened in 1969. It was the first prominent international conference on environmental issues. It was followed by the 1972 Stockholm Conference. Maurice Strong's "success in increasing foreign aid brought him to the attention of U Thant, the secretary general of the United Nations at the time, who selected him to convene the 1972 Stockholm conference" (Roberts, 2015).

U Thant's appointment of Strong required approval by the Canadian government. Pierre Elliott Trudeau, Canadian Prime Minister from the Liberal Party, authorized Strong's appointment. Strong began work at the UN headquarters in New York. He served as both Secretary General of the 1972 Stockholm Conference and UN Under-secretary General in charge of environmental affairs.

Strong applied his "consummate diplomatic skills to obtain the support of the developing countries, who were extremely skeptical about environmental issues" (Manitou-Strong, 2020). The 1972 Stockholm Conference "adopted a Declaration of Principles and Action Plan to deal with global environmental issues. It put the environmental issue on the international agenda and confirmed its close link with development. The Stockholm Conference moved into the history books as a major landmark, launching a new era of international environmental diplomacy" (Manitou-Strong, 2020).

Strong expressed his concern for the human environment and the need for global ecology in his opening remarks at the 1972 Stockholm Conference (Manitou-Stockholm, 2020). He began his remarks by saying that "We have made a global decision of immeasurable importance to which this meeting testifies: we have determined that we must control and harness the forces which we have ourselves created."

Strong was concerned about the possible occurrence of an environmental catastrophe although he did not think one was imminent. He stated that "Our whole work, our whole dedication is surely towards the idea of a durable and habitable planet." He was aware that most people around the world did not live sustainably. He asked, "can the great venture of human destiny be carried safely into a new century if our work is left in this condition? I, for one, do not believe it can."

The 1972 Stockholm Conference, according to Strong, was the human environment. He believed that the term "human environment" should be interpreted broadly: "the human environment impinges upon the entire condition of man, and cannot be seen in isolation from war, and poverty, injustice and discrimination."

Strong said that "all nations must accept responsibility for the consequences of their own actions on environments outside their borders." He emphasized that "Our major motivation in gathering here is to consider recommendations, which can only be translated into action by international agreement. By far the major part of the burden of environmental management falls, however, upon national governments operating as sovereign national states."

Strong argued that climate issues should be resolved using an incremental approach. He said, "I cannot stress too strongly, the central importance of accepting this notion of ongoing process, of continuity, of adaptation, of steady evolution, in perception, in organization, in decision making and in action to protect and enhance the human environment."

Strong said that "Many of the fundamental environmental problems of the developing countries derive from their very poverty and lack of resources and, in some cases, from inappropriate forms of development."

Strong did not support unlimited growth. He described the type of growth he supported: "People must have access to more, not fewer opportunities to express their creative drives. But these can only be provided within a total system in which man's activities are in dynamic harmony with the natural order. To achieve this, we must control and redirect our processes of growth. We must rethink our concepts of the basic purposes of growth. We must see it in terms of enriching the lives and enlarging the opportunities of all mankind. And if this is so, it follows that it is the more wealthy societies — the privileged minority of mankind — which will have to make the most profound, even revolutionary changes in attitudes and values."

Strong recognized that change was likely to be gradual and that developed nations would need to rethink how they managed their natural resources. "The overall global goal of the United Nations environmental program," he said, "must be to arrest the deterioration and begin the enhancement of the human environment." He reminded his audience that "The basic task of this conference is to build in the minds of men the new vision of the larger, richer future which our collective will and energies can shape for all mankind, to build a program of concerted action which will make an important first step towards the realization for this vision; to build the new vehicle of international cooperation that will enable us to continue the long journey towards that creative and dynamic harmony between man and nature that will provide the optimum environment for human life on Planet Earth."

The decision to establish the United Nations Environment Program (UNEP) in December 1972 by the UN General Assembly was the key outcome of the 1972 Stockholm Conference. The role of UNEP was to organize, promote, and catalyze action within the UN system. UNEP was not permitted to carry out or fund action.

Strong was selected as the first Executive Director of UNEP by the UN General Assembly. UNEP was headquartered in Nairobi, Kenya. It became the first UN agency to be headquartered in a developing country.

Canadian Prime Minister Pierre Trudeau invited Strong "to head the newly created national oil company, PetroCanada." Strong accepted the position and returned to Canada in 1976. He later "became Chairman of the Canada Development Investment Corporation, the holding company for some of Canada's principal government-owned corporations" (Manitou-Strong, 2020).

Strong served as a member of the Brundtland World Commission on Environment and Development (WCED). The WCED was presided over by Norwegian politician Gro Harlem Brundtland. The work of the Brundtland Commission was summarized in *Our Common Future* (WCED, 1987).

9.5 Agenda 21 and the 1992 Rio Conference

A new global environmental conference was convened on the 20th anniversary of the 1972 Stockholm Conference. The United Nations Environment Programme (UNEP) reported that "In June 1992 (Maurice Strong) was asked to lead another landmark meeting: the UN Conference on Environment and Development — also known as the

Earth Summit — it was held in Rio de Janeiro, Brazil" (UNEP-Strong, 2023). The formulation of Agenda 21, a program of action for attaining sustainable development in the 21st century, was a significant result of the United Nations Conference on Environment and Development (UNCED), now known as the 1992 Rio Conference.

Strong updated his vision of the future in his opening statement to the 1992 Rio Conference. He welcomed his audience and then acknowledged that "despite significant progress made since 1972 in many areas, the hopes ignited at Stockholm remain largely unfulfilled" (Manitou-Rio, 2020). He highlighted observations reported by Gro Harlem Brundtland's report *Our Common Future* to the World Commission on Environment and Development (WCED, 1987): "the environment, natural resources and life-support systems of our planet have continued to deteriorate, while global risks like those of climate change and ozone depletion have become more immediate and acute." Strong cautioned that "all the environmental deterioration and risks we have experienced to date have occurred at levels of population and human activity that are much less than they will be in the period ahead."

Strong specified the main concerns of the 1992 Rio Conference as "patterns of production and consumption in the industrial world that are undermining the Earth's life-support systems; the explosive increase in population ... deepening disparities between rich and poor that leave 75 per cent of humanity struggling to live; and an economic system that takes no account of ecological costs or damage."

"Population must be stabilized, and rapidly," Strong said. "If we do not do it, nature will, and much more brutally." Strong then connected Gross Domestic Product (GDP) to population control by observing that "world GDP increased by $20 trillion. Yet 15 percent of the increase accrued to developing countries" during the 20 years between the 1972 Stockholm Conference and the 1992 Rio Conference. Over 70% of the increase in GDP went to developed countries. "This is the other part of the population problem," he said. "The fact that every child born in the developed world consumes 20 to 30 times the resources of the planet than a third world child." He concluded that "The wasteful and destructive lifestyles of the rich cannot be maintained at the cost of the lives and livelihoods of the poor, and of nature."

Strong called on all governments participating in the 1992 Rio Conference to prioritize acceptance and implementation of the decisions made at the conference into policies and practices. He identified Agenda

21 as "a framework for the systemic, co-operative action required to effect the transition to sustainable development." He said "The issue of new and additional financial resources to enable developing countries to implement Agenda 21 is crucial and pervasive. This, more than any other issue, will clearly test the degree of political will and commitment of all countries to the fundamental purposes and goals of this Earth Summit."

According to Strong, "poverty and hunger persist in a world never better able to eliminate them. This is surely a denial of the moral and ethical basis of our civilization as well as a threat to its survival." He went on to say that "Agenda 21 measures for eradication of poverty and the economic enfranchisement of the poor provide the basis for a new worldwide war on poverty. Indeed, I urge you to adopt the eradication of poverty as a priority objective for the world community as we move into the 21st century."

Strong observed that "Nowhere is efficiency more important than in the use of energy. The transition to a more energy-efficient economy that weans us off our overdependence on fossil fuels is imperative to the achievement of sustainable development."

In conclusion, Strong said that "It is an exhilarating challenge to erase the barriers that have separated us in the past, to join in the global partnership that will enable us to survive in a more secure and hospitable world. The industrialized world cannot escape its primary responsibility to lead the way in establishing this partnership and making it work. Up to now, the damage inflicted on our planet has been done largely inadvertently. We now know what we are doing. We have lost our innocence. It would be more than irresponsible to continue down this path."

He completed his remarks by telling his audience that "The road beyond Rio will be a long and difficult one; but it will also be a journey of renewed hope, of excitement, challenge and opportunity, leading as we move into the 21st century to the dawning of a new world in which the hopes and aspirations of all the world's children for a more secure and hospitable future can be fulfilled. This unprecedented responsibility is in your hands."

The acceptance of Agenda 21, a non-binding agreement that presented a plan for attaining sustainable development, was one of the most significant steps made at the 1992 Earth Summit in Rio de Janeiro. President George H.W. Bush of the United States signed the Agenda 21 accord while attending the conference. In an attempt to facilitate sustainable development throughout the world, Agenda 21 encouraged

- countries with the largest populations to reduce population growth,
- countries with the largest consumption to reduce consumption, and
- the wealthiest countries to increase financial and technological aid to poorer countries.

In its obituary to Maurice Strong, *The New York Times* reported that "donor nations agreed to provide $7 billion in aid to poorer ones, the sum was far short of the $70 billion that the United Nations said was needed annually" (Roberts, 2015).

A few significant connections between Maurice Strong and Agenda 21 are outlined in Figure 9.1.

9.6 The Earth Charter Initiative

The Earth Charter Initiative was launched in 1994 by Maurice Strong and Mikhail Gorbachev, the eighth and final leader of the Soviet Union. Gorbachev led the Soviet Union from 1985 to its dissolution in 1991. The Earth Charter was supposed to be finished at the 1992 Earth Summit. According to Maurice Strong, "the Earth Charter would set out the basic principles for the conduct of people and nations toward each other and the Earth to ensure our common future" (Strong, 2000, p. 202). The Preamble to the Earth Charter is presented in Table 9.1 (Earth Charter, 2023).

"The real goal of the Earth Charter," according to Strong, "is that it will in fact become like the Ten Commandments, like the Universal

Maurice Strong
(1929-2015)
↓
United Nations
(Founded 1945)
↓
Stockholm Conference
(1972)
↓
UNEP
(Founded 1972)
↓
Rio Conference: Agenda 21
(1992)

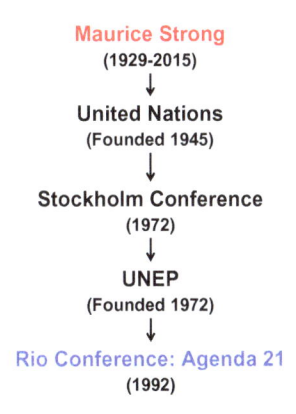

Figure 9.1 Connections from Maurice Strong to Agenda 21

Table 9.1 Preamble to the Earth Charter

"We stand at a critical moment in Earth's history, a time when humanity must choose its future. As the world becomes increasingly interdependent and fragile, the future at once holds great peril and great promise. To move forward, we must recognize that in the midst of a magnificent diversity of cultures and life forms, we are one human family and one Earth community with a common destiny. We must join together to bring forth a sustainable global society founded on respect for nature, universal human rights, economic justice, and a culture of peace. Towards this end, it is imperative that we, the peoples of Earth, declare our responsibility to one another, to the greater community of life, and to future generations."

Declaration of Human Rights. It will become a symbol of the aspirations and commitments of people everywhere. And that is where the political influence, where the long-term results of the Earth Charter will really come" (Strong, 1998).

The Earth Charter International Secretariat and Council paid respect to Maurice Strong after his passing in 2015. It said that "As a member of the Brundtland Commission and Secretary General of the Earth Summit in 1992, Mr. Strong took on the commitment to carry the idea of an Earth Charter forward (which was a recommendation made in the Brundtland Commission Report and according to him an unfinished piece of business of the Rio Earth Summit). Therefore, in 1994, together with Mikhail Gorbachev he launched the Earth Charter Initiative and became the co-Chair of the International Commission" (ECI-Strong, 2015).

9.7 Strong's Later Years

In 2000, Maurice Strong began his book *Where on Earth Are We Going?* With a scenario that illustrated what the world could look like in 30 years if anthropogenic climate change was not mitigated (Strong, 2000). According to Strong, the world's population might be reduced by two-thirds by the year 2031 because of climate change, environmental catastrophes, water scarcity, the resurgence of previously controlled diseases, political unrest, and the breakdown of law and order.

Strong's book served as an appeal for leaders of industrial, governmental, and environmental organizations to find solutions to global problems in the new millennium. "The time has come," he said, "when we need to act both globally and locally, and that requires the cooperation of all of us, from individuals to grassroots groups to business, governments and supranational organizations" (Strong, 2000, p. 5).

Strong knew that he had been associated with a movement to create a global government. He thought about transforming the UN into a global government as one possibility. To clarify his views, he maintained that a global government "is not necessary, not feasible, and not desirable" (Strong, 2000, p. 319). Twenty-two pages later he expressed an apparently contradictory view that "the codification, administration and enforcement of international law must become one of the principal functions of the United Nations in the period ahead" (Strong, 2000, p. 341).

Strong did not believe that the world should become an anarchical world without systems or rules: "A chaotic world would pose equal or even greater danger. The challenge is to strike a balance so that the management of global affairs is responsive to the interests of all people in a secure and sustainable future. Such management must be guided by basic human values and make global organization conform to the reality of global diversity" (Strong, 2000, p. 319).

Global institutions like the UN could provide the basic components "of an improved system of international agreements and international law and more streamlined international organizations to service and support the cooperation among governments and other key actors that will be required" (Strong, 2000, p. 321). Strong believed that the rule of law was essential to the functioning of societies on the national and global scale. He realized, however, that the "single greatest weakness of the existing international legal regime is the almost total lack of a capacity for enforcement" (Strong, 2000, p. 342).

Strong left his position as UN representative to North Korea in 2005 "after Tongsun Park, a South Korean with a scandalous past, was found to have been an unregistered lobbyist for Iraq in the United Nations oil-for-food program and to have invested $1 million in a company controlled by Mr. Strong. Mr. Strong was cleared of any involvement in the scandal" (Roberts, 2015). After 2005, Strong spent a significant amount of time in China working as an honorary professor at Beijing's Peking University. Maurice Strong passed away in 2015.

Chapter 10

Obama Picks Up the Environmental Baton

Maurice Strong left a leadership vacuum in the global environmental movement when he left the United Nations in 2005. The vacuum could have been filled by former United States Vice President Al Gore if he would have won the Presidency in the 2000 election. Instead, George W. Bush won the 2000 election and sent Gore to the private sector.

Gore created a presentation for the public that was designed to increase awareness about the risks of global warming. He then helped write and present the 2006 documentary "An Inconvenient Truth" that chronicled his efforts to inform the public about climate change. The movie received two Academy Awards from the American Academy of Motion Picture Arts and Sciences in 2007 — one for Best Documentary Feature and one for Best Original Song (Gore-Oscars, 2007). Furthermore, Gore and the Intergovernmental Panel on Climate Change (IPCC) shared the 2007 Nobel Peace Prize "for their efforts to build up and disseminate greater knowledge about man-made climate change, and to lay the foundations for the measures that are needed to counteract such change" (Gore-Nobel Prize, 2007).

The leadership vacuum in global environmentalism was filled in 2008 when Democrat Senator Barack Obama defeated Republican Senator John McCain to win the United States presidential election. Gore recognized that Obama's election answered the question of who would address problems arising from anthropogenic climate change. He said that "The turning point came in 2009 ... with the inauguration of a new President (Obama) in the United States, who immediately shifted priorities to focus on building the foundation for a new low-carbon economy" (Gore, 2009, p. 394). The following section summarizes Obama's climate initiatives.

10.1 Obama's Climate Initiatives

Barack Obama was inaugurated as President of the United States in January 2009 and subsequently appointed several environmentalists to lead his departments. Figure 10.1 shows some key Obama–Biden Administration officials and their government affiliations.

Carol Martha Browner, an attorney and environmentalist, led Obama's White House Office of Energy and Climate Change Policy from 2009 until 2011. The Domestic Policy Council took over the office's work on climate and energy after it was abolished as a standing White House agency in 2011.

Obama's scientific and technology adviser was John P. Holdren. Holdren held degrees in theoretical plasma physics and aerospace engineering. Obama served two terms as President. During this period from 2009 to 2017, Holdren was Director of the White House Office of Science and Technology Policy. He also served as Assistant to the President for Science and Technology and Co-Chair of the President's Council of Advisors on Science and Technology (PCAST).

Chemical engineer Lisa Jackson was an authority in air quality and environmental health. She led the United States Environmental Protection Agency (EPA) as administrator during Obama's first term (2009–2013).

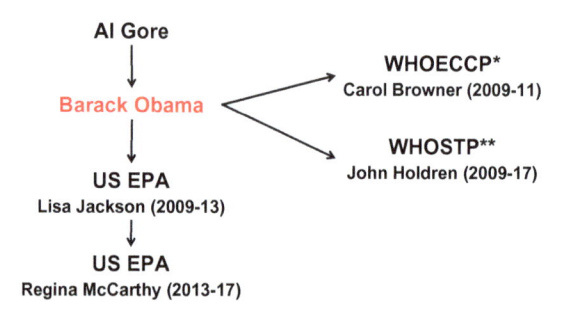

Figure 10.1 Some Climate Policy Players in the Obama–Biden Administration
*White House Office of Energy and Climate Change Policy
**White House Office of Science and Technology Policy

Table 10.1 Key Climate Change Areas during the Obama–Biden Administration

1	Decrease carbon pollution
2	Expand the carbon-free economy
3	Lead global efforts to mitigate anthropogenic climate change
4	Protect climate, air, and water
5	Improve energy use efficiency
6	Protect natural resources

Regina McCarthy took over Jackson's EPA responsibilities during Obama's second term (2013–2017).

During Obama's two terms as president, his administration implemented several energy and climate measures. According to the Obama White House Archives (OWHA), President Obama made "a historic commitment to protecting the environment and addressing the impacts of climate change" (OWHA Climate, 2023). In an introductory statement, the OWHA document says that "President Obama believes that no challenge poses a greater threat to our children, our planet, and future generations than climate change — and that no other country on Earth is better equipped to lead the world towards a solution. That's why under President Obama's leadership, the United States has done more to combat climate change than ever before, while growing the economy."

The Obama–Biden Administration sought to protect the environment and address possible effects of climate change in the six key areas listed in Table 10.1. Many of the actions taken by the Obama–Biden Administration were not enacted as laws but as regulations or executive orders. For example, the Paris Climate Agreement was signed by the Executive Branch but did not have the authority of a Treaty because it was not approved by the United States Senate. A new president had the authority to rescind or change executive actions without Congressional approval if the actions were not enacted into law by Congress.

10.2 Trump Disrupts Obama's Climate Agenda

The Obama–Biden Administration favored many climate and energy policies that were rejected by the succeeding Trump Administration.

After being inaugurated in January 2017, President Donald Trump and his administration began repealing Obama-era regulations and agreements. For example, in June 2017, the Trump Administration withdrew the United States from the COP21 Paris Climate Agreement, citing concerns that the agreement would have a greater detrimental impact on the nation's economy than a beneficial impact on the global environment. Policy differences between the two administrations illustrate the observation that political parties can have competing energy visions.

The past two decades of United States energy policy demonstrate the impact of competing energy visions. The political right argues that the role of the federal government should be minimal in the development of carbon-free energy. They believe that the free market should be used to develop appropriate technology and determine the length of time needed to complete the transition to a new energy portfolio. The political left, on the other hand, says that the environment is being harmed because the price of different forms of energy in the free market does not adequately account for the cost of pollution and greenhouse gas emissions.

Governments can impose environmental costs on different energy forms by taxing and regulating industries. The imposition of environmental costs increases the cost of the resource. The political left argues that this makes the cost more equitable. Government (taxpayer) funds can also be used to support emerging carbon-free technologies which private enterprises are unwilling to fund. The political right responded with the argument that government intervention in the free market will distort prices and could artificially inflate or reduce prices in an unsustainable manner. The finite resources available to the government might be better used elsewhere. Furthermore, the involvement of the government raises the possibility of fraud, corruption, and political favoritism.

The War on Coal illustrates the partisan divide between political parties on energy policy. Scientists have observed that more greenhouse gas is emitted by the combustion of coal than the combustion of other fossil fuels. The Obama–Biden Administration provided resources to expand the carbon-free economy by subsidizing renewable energy. New federal resources were also included to moderate public concerns about job losses. This effort was needed to mitigate a publicly perceived war on coal that arose from a November 7, 2008, interview between then presidential candidate Obama and the San Francisco Chronicle. In the interview, Obama said "If somebody wants to build a coal powered plant, they can, it's just that it will bankrupt them because they are going to be charged a

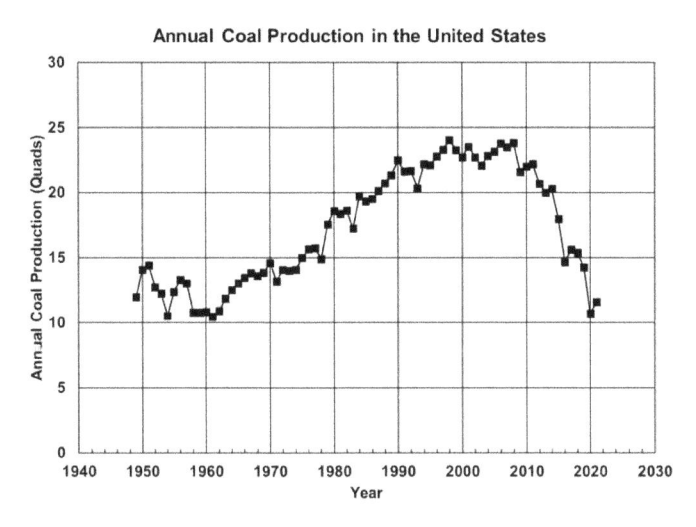

Figure 10.2 Annual Coal Production in the United States (US EIA MER Table 1.2, 2022)

huge sum for all that greenhouse gas that's being emitted." Obama stated in the interview that he was willing to support the implementation of clean coal technologies and sequestration of greenhouse gases, but his animosity towards coal combustion became characterized as a war on coal.

The decrease in coal production from 2009 to 2016, the duration of Obama's two terms, is depicted in Figure 10.2. Many coal-fired power facilities were decommissioned or converted to natural gas-fired power facilities. The figure shows that coal production increased during the first 2 years of Donald Trump's 2017–2021 presidential term, followed by a decline during the COVID-19 pandemic.

Point to Ponder: What is the COVID-19 Pandemic?

The COVID-19 virus appears to have spread from Wuhan, China to other parts of the world in late 2019 and early 2020. In response to the resulting global pandemic, the economies of several nations were partially shut down and travel was curtailed to slow the spread of the virus. The result was a decrease in both energy production and consumption.

10.3 Biden Resurrects Obama's Climate Agenda

Donald Trump favored an energy production strategy that relied on nuclear fission, fossil fuels, and alternative energy sources. This strategy is known as an "all of the above" strategy. Trump was unable to revive coal production in the early 2020s because he lost his 2020 bid for election to a second consecutive term. He is seeking a second term as President by running as a Republican candidate in 2024 using the fossil fuel friendly slogan 'Drill, baby drill.'

During the 2020 campaign against Trump, Joe Biden said he would resume many of the policies that were part of the Obama–Biden Administration. This created an interesting dynamic, especially after a December 1, 2020, interview with talk show host Stephen Colbert. Colbert asked Obama if he missed being president when he looked at media coverage of the Trump administration. Obama had already completed two terms as president and was prohibited from serving a third term. According to commentator C. Douglas Golden, "Obama mentioned that there were plenty of people who wanted him to have a third term." Golden then quoted Obama's reply to Colbert: "If I could make an arrangement where I had a stand-in, a frontman or frontwoman, and they had an earpiece in, and I was just in my basement in my sweats, looking through the stuff, then deliver the lines, but somebody else was doing all the talking — I'd be fine with that.

"Because I found the work fascinating. Even on my worst days, I found puzzling out these big, complicated, difficult issues — especially if you're working with some great people — to be professionally, really satisfying. But I do not miss having to wear a tie every day" (Golden, 2020).

The idea that Obama had found his front man for a 'third term' in the person of aging Joe Biden was reinforced when Biden announced his White House staff and Cabinet members. National Public Radio (NPR) reported that then President-elect Joe Biden's picks for his Cabinet and White House team would be "the most diverse team in history. But in his picks so far, there is one thing that most of his team will have in common: service in the Obama administration" (Ordoñez, 2020). Ideologically, there seemed to be little difference between a Biden presidency and the Obama presidency.

The view that Obama is still involved in governing the United States is embraced by contributors to *Barack Obama's True Legacy*. In the

Epilogue, Robert Spencer wrote that "Barack Obama's third term has begun, in the guise of the Biden/Harris administration. The defeat of Donald Trump and the restoration of (Obama's) political establishment's power makes this present book not just a record of the recent past, but a warning for the Biden years and their aftermath" (Spencer in Glazov, 2023, p. 248).

Obama's involvement in Biden's presidency surfaced again when Biden issued an executive order to regulate Artificial Intelligence (AI) on October 30, 2023 (AI Biden, 2023; Villalovis, 2023). AI algorithms can be used to control access to information. For example, rules governing search algorithms can specify the legitimacy of information and censor any information that is classified as incorrect or misleading information (misinformation), or intentionally deceptive information (disinformation).

In January 2021, Democrat Joe Biden, Obama's former Vice President, was inaugurated President. Biden had already promised to revive the Obama–Biden coal policy in the first presidential debate between Biden and Republican Donald Trump on September 29, 2020, in Cleveland, Ohio. Biden said "Nobody's going to build another coal-fired plant in America. No one is going to build another oil-fired plant in America. They're going to move to renewable energy" (Boyd, 2020).

Approximately 2 years into his presidency, in the 2022 midterm election season, Biden reaffirmed his position on coal: "we're going to be shutting these (coal) plants down all across America and having wind and solar." The Obama–Biden War on Coal began during the Obama Administration, was interrupted by the Trump administration, and then resumed by the Biden Administration. It is an example of the politization of American energy policy (Frazin, 2022).

Part 4

Achieving Global Governance

Chapter 11

Weaponizing the Privileged Minority

An explanation of how climate change can be weaponized to empower a global government was presented in Part 3. Here we discuss how global governance is being achieved by weaponizing institutions and events. The emerging global government is being used to attack national sovereignty and change the world order.

Chapter 11 shows how the privileged minority can use its assets to destabilize sovereign states. The weaponization of national governments in the struggle to achieve global governance is described in Chapter 12. Human conflicts provide additional opportunities for globalists to seize control of sovereign states, as discussed in Chapter 13. An example, the COVID-19 pandemic, is discussed in Chapter 14 as an example of how quickly human rights can be lost to achieve security. With this background, a modern model of changing world orders and a selection of possible scenarios for a world order that might emerge after the energy transition are discussed in Chapter 15.

11.1 The Privileged Minority and Globalization

Maurice Strong was concerned that his plea for action to address anthropogenic climate change was not being supported by the "privileged minority" (Strong, 2000, p. 28). In this context, the privileged minority is a coterie of people who control considerable wealth, power, or influence. Some members of the privileged minority can exercise power in one or more arenas, such as financial, industrial, or political.

Strong believed the privileged minority was not experiencing sufficient hardship to understand and appreciate the need to mitigate anthropogenic climate change. He was able to identify a few exceptions among the privileged minority that were in a position to assist through non-profit and volunteer organizations: "Particularly noteworthy are George Soros, who donates hundreds of millions of dollars a year through his Open Society Foundation, largely in the countries of the former Soviet Union, and the billionaire media genius Ted Turner, who made the largest single charitable contribution in US history by committing $1 billion to support United Nations programs and activities. His generosity has since been topped by the computer software king Bill Gates, the first person ever to have his personal net worth reach $100 billion. They follow in the tradition of the great philanthropists of the past – notably the Rockefeller family, which continues in its current generation to set a remarkable example of enlightened and innovative philanthropic leadership" (Strong, 2000, p. 28). In the next section, we illustrate the loosely connected international network of the privileged minority that is attempting to alter the current world order by highlighting one funding source — the Rockefeller family.

11.2 Amassing the Rockefeller Fortune

The Rockefeller fortune was created by John Davison Rockefeller (1839–1937) (Tarbell, 1904; Yergin, 1992; BOE Standard Oil, 2023). Together with Maurice B. Clark and Samuel Andrews, he founded an oil refinery in Cleveland, Ohio, in 1863. Rockefeller bought out Clark in 1865, and invited Henry M. Flagler to join the firm of Rockefeller, Andrews, and Flagler as a partner in 1867. Their oil refinery was acquired by Standard Oil Company, a business that was established in 1870 and headquartered in Cleveland. By 1880, the company controlled over 90% of oil refining in the United States. It achieved dominance in oil refining by eliminating rivals, merging with other businesses, and using its size and efficiency to obtain favorable railroad rebates, or discounts, on railway freight prices.

A holding company called Standard Oil Trust was established in 1882 by Rockefeller and his associates by combining companies in production, refining, and marketing. According to Daniel Yergin, "A board of trustees was set up, and in the hands of those trustees was placed the stock of all the entities controlled by Standard Oil. Shares in turn were issued in the trust; out of the 700,000 total shares, Rockefeller held 191,700 and

Flagler, next, had 60,000" (Yergin, 1992, p. 43). In addition, "separate Standard Oil organizations were set up in each state to control the entities in those states" (Yergin, 1992, p. 45). Standard Oil of New Jersey was one such organization. The headquarters of Standard Oil Trust moved to New York in 1885.

Congress was concerned that Standard Oil Trust would be a monopoly due to its size and scope. John Sherman, a United States Senator from Ohio, proposed an antitrust law that would permit the federal government to break up enterprises that impede competition. The Sherman Antitrust Act was enacted by the US Congress in 1890.

The State of Ohio used the antitrust act in 1892 to file suit against Rockefeller and Standard Oil. The Ohio Supreme Court ordered the Trust to break up that same year. The decision was appealed by the Trust. The appeal allowed the Trust to continue operating from its New York headquarters. More than a decade later, in 1906, the administration of then President Theodore Roosevelt sued Standard Oil for conspiring to restrain trade under the Sherman Antitrust Act (Yergin, 1992, p. 108).

Standard Oil was forced to breakup into many smaller companies in 1911 by the United States Department of Justice. J.D. Rockefeller became the richest man in the world following the breakup of Standard Oil Trust (Yergin, 1992, p. 113). Standard Oil of New Jersey was one of the companies that was formed by the breakup of Standard Oil Trust. Over the years it became part of Exxon and then ExxonMobil. Significant events in the evolution of Standard Oil are highlighted in Figure 11.1.

Cleveland Refinery ← **John D. Rockefeller**
(Founded 1863) (1839-1937)
↓
Standard Oil of Ohio
(Founded 1870, HQ in Ohio)
↓
Standard Oil Trust
(Founded 1882, moved HQ to NY in 1885)
↓
Sherman Antitrust Act → **J.D. Rockefeller Retires**
(Signed into Law 1890) (from SONJ, 1896)
↓
Breakup of Standard Oil → **J.D. Rockefeller became**
(1911) **richest man in world**

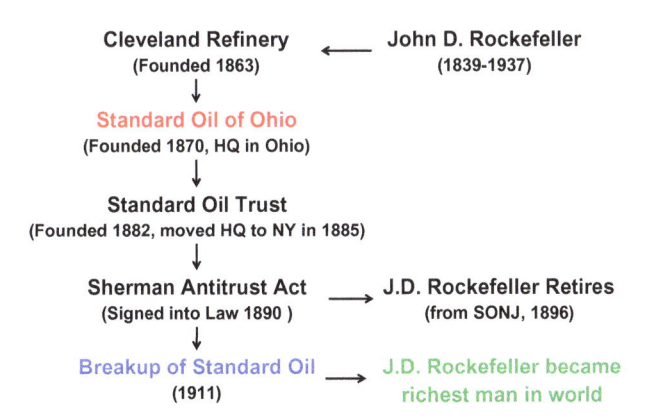

Figure 11.1 The Rise and Fall of Standard Oil

11.3 The Rockefeller Foundation

The Rockefeller Foundation was founded in New York State by John D. Rockefeller, his only son John D. Rockefeller, Jr. (JDR Jr.), and Frederick Taylor Gates. Gates, a former Baptist Minister, became an advisor to J.D. Rockefeller, Sr. (JDR Sr.) in the oil and gas industry, and philanthropy. The New York State Legislature accepted the Rockefeller Foundation charter on May 14, 1913. Figure 11.2 displays the lineage between the Rockefellers who founded the Rockefeller Foundation and a selection of Rockefeller family members that significantly impacted the Rockefeller Foundation in later years.

The Rockefeller Foundation charter was accepted by the New York State Legislature on May 14, 1913. Raymond B. Fosdick wrote a history of the Rockefeller Foundation from the date its charter was accepted until 1948 when Fosdick's 12-year tenure as President of the Rockefeller Foundation ended (Fosdick, 1952). Fosdick had a thorough understanding of the Foundation's history during this period.

Fosdick attributed the idea to establish a Foundation to Frederick Taylor Gates. In a 1905 letter to John D. Rockefeller, Gates wrote that "It seems to me that either you and those who live now must determine what shall be the ultimate use of this vast fortune, or at the close of a few lives now in being it must simply pass into the unknown, like some other great fortunes, with unmeasured and perhaps sinister possibilities" (Fosdick, 1952, p. 14). JDR Sr. had many discussions about the letter with his son JDR Jr. and Gates. They were interested in beginning a trust that would

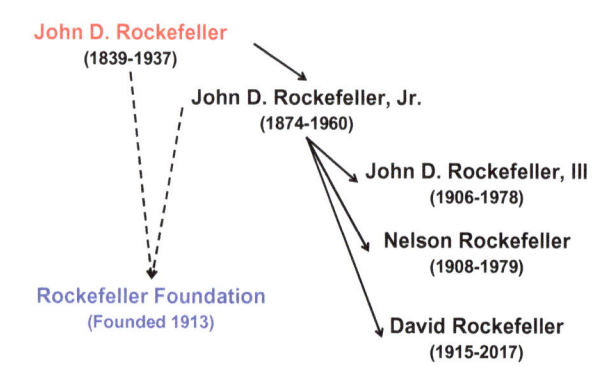

Figure 11.2 Rockefeller Lineage and Founders of the Rockefeller Foundation

contain "considerable sums of money to be devoted to philanthropy, education, science, and religion" (Fosdick, 1952, p. 15).

JDR Jr. started working for his father after earning his degree from Brown University in 1897. According to JDR Jr., "if I was going to learn to help Father in the care of his affairs, the sooner my apprenticeship under his guidance began, the better" (Fosdick, 1952, p. 3). Fosdick observed that "Few father-and-son relationships have been characterized by more genuine trust or deeper affection. For 40 years they worked closely and intimately together" (Fosdick, 1952, p. 3). The three-cornered partnership between Gates, JDR Sr. and JDR Jr. "was responsible for a group of foundations to which Mr. Rockefeller (JDR Sr.) contributed nearly half a billion dollars" (Fosdick, 1952, p. 3).

Gates cautioned JDR Sr. that "Your fortune is rolling up, rolling up like an avalanche! You must keep up with it! You must distribute it faster than it grows! If you do not, it will crush you and your children and your children's children!" (Fosdick, 1952, p. 3). Fosdick claimed that JDR Sr. was influenced by Gates' warning and an essay written in 1889 by steel industrialist and philanthropist Andrew Carnegie (1835–1919). In his essay, Carnegie said that "the man who dies leaving behind him millions of available wealth, which was his to administer during life, will pass away 'unwept, unhonored, and unsung'… The man who dies thus rich dies disgraced." JDR Sr. concluded that "a man should make all he can and give all he can" (Fosdick, 1952, p. 6).

JDR Sr. established a set of criteria for contributions in the 1880s: "His money should be given to work already organized and of proven worth; it should be work of a continuing character which would not disappear when his gifts were withdrawn; the contributions, where possible, should be made on conditional terms so as to stimulate contributions by others; and finally… his money should make for strength rather than weakness and should develop in the beneficiary a spirit of independence and self-reliance" (Fosdick, 1952, p. 6).

JDR Sr. executed a deed of trust in 1909 that transferred shares of Standard Oil Company of New Jersey stock worth approximately $50,000,000 to three trustees: his son JDR Jr., his son-in-law Harold McCormick, and Frederick T. Gates. The trust was known as The Rockefeller Foundation. Its trustees began the task of obtaining a corporate charter.

In March 1910, a bill to incorporate The Rockefeller Foundation was introduced in the United States Senate. Congress did not act on the bill for years because some members were concerned that JDR Sr. was implementing a plan to perpetuate his wealth. According to Fosdick, JDR Sr. was seeking government control to make sure that the funds would be used for public good.

Congress had not passed the bill by the time it adjourned in 1913. JDR Sr.'s advisers called on the New York State Legislature in Albany to incorporate The Rockefeller Foundation. The Legislature incorporated The Rockefeller Foundation in 1913, a few months after the adjournment of Congress. The Rockefeller Foundation was created during the presidency of Woodrow Wilson, a leader of the Progressive Movement in the United States. Wilson served as President from 1913 to 1921. The mission of the Rockefeller Foundation was "to promote the well-being of mankind throughout the world" (Fosdick, 1952, p. 20).

Forty years after its incorporation, Fosdick wrote that "in a deep and ultimate sense, it is still one world, one human race, one common destiny. That was the high faith that lay behind the creation of the Foundation, and on that faith the future must depend" (Fosdick, 1952, p. 288). He expressed this sentiment at a time when World War II had ended, and the Cold War with the Soviet Union was beginning. Fosdick said that humanity had to come to terms with the "grim necessities which today's problem of security brings to all of us" (Fosdick, 1952, p. 288).

According to the Rockefeller Foundation, "Since 1913, The Rockefeller Foundation has pursued our mission to promote the well-being of humanity around the world by breaking down the barriers that limit who can be healthy, empowered, nourished, well off, secure, and free." Today, the Rockefeller Foundation identifies itself as "a philanthropic foundation that promotes the well-being of humanity by finding and scaling solutions to advance opportunity and reverse the climate crisis" (RF-Mission, 2023). In the next section, we review some of the connections between the Rockefeller Foundation and climate activism.

11.4 David Rockefeller

Climate activist Maurice Strong recalled meeting David Rockefeller, the son of John D. Rockefeller, Jr., while Strong was a young man in New York seeking a position with the United Nations. Strong and Rockefeller participated in or worked with a number of the same organizations and initiatives during their respective careers. Some international

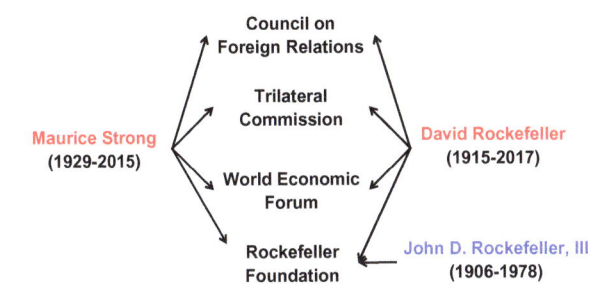

Figure 11.3 A Few Connections between Maurice Strong and Members of the Rockefeller Family

organizations that connect Maurice Strong and banker David Rockefeller are highlighted in Figure 11.3. The figure also shows connections that connect Strong with David's oldest brother John Davison Rockefeller, III (JDR III). JDR III played a leading role in the Rockefeller Foundation. In this section, we review some aspects of David's life that provide insight into the modern view of internationalism held by the privileged minority.

11.4.1 *Early Years*

Born in 1915, David Rockefeller was the son of John D. Rockefeller, Jr. (JDR Jr.) and Abigail Green Aldrich Rockefeller (1874–1948). Abigail's brother was lawyer Winthrop Aldrich (1885–1974). Winthrop was appointed Chairman of Chase Bank following a stock selling scandal in 1933. Two structural banking reforms were enacted by Congress following the scandal: the Glass-Stegal Act, which distinguished between commercial and investment banking, and the Security Act, which required corporations to register their stock and provide financial reports to the Securities and Exchange Commission (SEC) (Rockefeller, 2002, p. 125).

The Great Depression began in 1929 following the Wall Street collapse. David was a student at Harvard during the Great Depression. He completed his undergraduate degree in 1936 and remained to pursue graduate-level studies in economics for an additional year. During his time at Harvard, David had the opportunity to learn from economists who were debating the role of government in economics.

British economist John Maynard Keynes (1883–1946) believed that an economy could be stimulated by government intervention. By contrast, free market economists were concerned that the free market would be replaced by Keynesian government intervention and result in permanent control of the economy by the government.

The London School of Economics (LSE) was founded by Fabian Society members Sidney and Beatrix Webb in 1895. David Rockefeller attended LSE after he left Harvard. He observed that "In those days the LSE was widely considered a hot bed of socialism and radicalism. Founded by the Webbs in the 1890s to help achieve their Fabian Socialist goal of a just society based on a more equal distribution of wealth, its walls had always given shelter to men and women who tested the limits of orthodoxy" (Rockefeller, 2002, p. 82).

According to David, political science professor Harold Laski "enthralled well-filled classrooms with his eloquent Marxist lectures" (Rockefeller, 2002, p. 82). Laski's work in the 1920s and 1930s contributed significantly to LSE's reputation. David "found the intellectual content of (Laski's) lectures superficial and often devious and deceptive. They seemed more propaganda than pedagogy" (Rockefeller, 2002, p. 82).

The relatively conservative economics department at LSE impressed David more favorably than Marxism. David was tutored by Friedrich von Hayek, an Austrian economist at LSE. Hayek's work on money, the business cycle, and capital theory in the 1930s and 1940s was awarded the Nobel Prize in Economics in 1974.

Hayek did not support Keynesian economics and its dependence on government intervention. Instead, he preferred free market competition. Hayek wrote *The Road to Serfdom* between 1940 and 1943 during the period when much of Europe was controlled by Nazi Germany and Fascist Italy. *The Road to Serfdom* expressed Hayek's concern that economic decision-making by government controlled central planning could replace a free market economy with a more tyrannical economy based on fascism or national socialism (Hayek, 1944).

In 1938, David left LSE to complete his graduate work at the University of Chicago. University of Chicago economists tended to believe that economic growth was more likely to be sustained by a free market and natural pricing mechanisms than government intervention. In 1939, David returned to New York to complete his doctoral dissertation. He received his doctorate in economics from the University of Chicago in

1940. Upon completion of his studies, he realized that he was "a pragmatist who recognizes the need for sound fiscal and monetary policies to achieve optimum economic growth" (Rockefeller, 2002, p. 92). He also concluded that society needed affordable safety nets.

11.4.2 *Philanthropy*

The philanthropic tradition of the Rockefellers "required that we be generous with our financial resources and involve ourselves actively in the affairs of our community and the nation" (Rockefeller, 2002, p. 145). David Rockefeller and his brothers advanced their interests in philanthropy by founding the Rockefeller Brothers Fund (RBF) in 1940. The brothers believed that the RBF would eventually make it possible for them "to work together and to forge a philanthropic philosophy that reflected our generation's values and objectives" (Rockefeller, 2002, p. 141).

David Rockefeller worked for New York Mayor Fiorello La Guardia (1882–1947) after completing his dissertation. When the United States entered World War II in 1941, Abigail, David's mother, made it clear that he had a 'duty' to participate in the war effort. In 1942, David joined the Army as a private.

David was promoted to corporal when he finished basic training and was assigned to Counter-Intelligence. He was accepted into Officer Candidate School (OCS) in 1943. Upon completion of OCS, he was commissioned as a second Lieutenant and assigned to Military Intelligence. He was deployed to Algiers as a Military Intelligence officer and was assigned the task of developing an intelligence network. One of the lessons he learned from his wartime experience was the value of networking: "I discovered the value of building contacts with well-placed individuals as a means of achieving concrete objectives" (Rockefeller, 2002, p. 122).

Chase National Bank hired David in 1946 after the war. David "began as an assistant manager, the lowest officer rank, in the foreign department at an annual salary of $3,500" (Rockefeller, 2002, p. 137). He noticed that the bank did not have a significant international presence and realized that this was an opportunity to expand the bank's foreign presence.

David traveled to Europe and other parts of the world in the late 1940s to find out where he could make the greatest impact. It was apparent to him that Latin America and the Caribbean were underserved markets. David developed a Caribbean strategy of establishing branch banks, acquiring local banks, and increasing loan activities. During this period, he first met a young Maurice Strong in New York.

The Chase branch system in the Caribbean "had emerged as the most dynamic part of our overseas operations" by the early 1950s (Rockefeller, 2002, p. 133). David wanted to extend the Caribbean strategy to other parts of the world and was on track to become Chair of Chase in 1969.

In the meantime, the brothers were funding the RBF by individually donating to RBF each year. RBF was not funded by an endowment during the first 12 years of its existence. By the 1950s, their "individual annual contributions to the RBF had grown to the point that we were able to support organizations which individual brothers had initiated or in which one of us had a special interest" (Rockefeller, 2002, p. 140).

11.4.3 *International Experience*

David Rockefeller developed a keen awareness of world events while traveling abroad in the 1930s and serving in another country during World War II. He enhanced his awareness "through active involvement with the Council on Foreign Relations, the Carnegie Endowment for International Peace, and International House of New York" (Rockefeller, 2002, p. 145). Furthermore, Chase established an International Advisory Committee (IAC) to improve the bank's connections with leaders around the world in the 1960s.

David retired in 1981and took over as IAC chairman (Rockefeller, 2002, pp. 208–209). He was a member of the Bilderberg group and wrote that "Bilderberg is really an intensely interesting annual discussion group that debates issues of significance to both Europeans and North Americans — without reaching consensus" (Rockefeller, 2002, p. 411). In 1985 he became chair of the Council on Foreign Relations.

11.4.4 *The Trilateral Commission*

The Trilateral Commission was established in 1973 by David Rockefeller, Zbigniew Brzezinski, and Jimmy Carter. The Commission's goal was to reach out to Japan, a leading economic power, in addition to North America and Europe about issues that, in his opinion, required cooperation from all three continents (Trilateral Commission Tribute, 2017).

Founding Director Zbigniew Brzezinski said in 2004 that trilateralism began as a strategy of seeking "trilateral cooperation between the three major democratic centers of economic and political power — between North America, Western Europe, and Japan" (Brzezinski, 2004). Brzezinski viewed trilateralism as "the key to global stability and progress; that therefore trans-Atlantic cooperation had to be wedded to trans-Pacific cooperation and that in the world at large democracy plus prosperity equal influence and power and that such power and influence should be harnessed for the common good, both for the sake of self-interest and of the moral imperative" (Brzezinski, 2004). Brzezinski concluded his address with his view that the Trilateral Commission was not only an "annual get-together of the democratic world's rich and influential — but as the institutional expression of a strategic vision that calls for action" (Brzezinski, 2004).

One of the issues considered by the Trilateral Commission was discussed in a 1972 report called *The Limits to Growth* (Meadows *et al.*, 1972). The report was published by The Club of Rome, an organization that seeks solutions to complex global issues. The report said that the finite supply of resources available to humanity could not sustain an exponential expansion of the human population.

In a Foreword to a Trilateral Commission report, David Rockefeller said that Zbigniew Brzezinski called *The Limits to Growth* a "pessimist manifesto" (Rockefeller, 1991). Vaclav Smil, an energy scholar, said *The Limits to Growth* was "easily the most widely publicized, and hence the most influential, forecast of the 1970s, if not the last one-third of the twentieth century" (Smil, 2003, p. 168). After studying *The Limits to Growth*, Smil concluded that the "report pretended to capture the intricate interactions of population, economy, natural resources, industrial production, and environmental pollution with less than 150 lines of simple equations using dubious assumptions to tie together sweeping categories of meaningless variables" (Smil, 2003, p. 169).

The Trilateral Commission published a study in 1991 that demonstrated the connection between the global economy and the environment (MacNeill *et al.*, 1991). The study was published prior to the 1992 Earth Summit Conference in Rio de Janeiro. North American Chairman of the Trilateral Commission David Rockefeller wrote the Foreword for the study, and Secretary General of the United Nations Conference on Environment and Development Maurice Strong wrote the Introduction.

11.4.5 *Internationalism in His Later Years*

Family members of David Rockefeller chose to reevaluate their shared goals following the turmoil of the Vietnam War years and the 1973 resignation of President Richard Nixon after the Watergate break-in. The Rockefellers realized that they shared a set of common desires: "to create a more just world that was free of racial intolerance and bigotry; to eliminate poverty; to improve education; and to figure out how the human race could survive without destroying the environment" (Rockefeller, 2002, p. 322).

David Rockefeller came to see that his extended family was also affected by the strife in society. John D. Rockefeller III, David's brother, dedicated his life to philanthropy. Normally reserved, JDR III rejected his aggressive younger brother Nelson's efforts to control the Rockefeller non-profit organizations. Nelson was a nationally recognized Republican politician who served as Gerald R. Ford's (1913–2006) Vice President from 1974 to 1977. Ford became President after Nixon resigned.

JDR III had served as chairman of the Rockefeller Foundation since the 1950s. He "viewed himself as the legitimate 'heir' of the Rockefeller tradition of philanthropy, which he also considered the core value of the Rockefeller family and the only activity that could over time hold family members together" (Rockefeller, 2002, p. 338). David seemed to agree with JDR III. He said that "Philanthropy was John's field, and he resented Nelson's assertions that it was he, rather than his older brother, who should guide the future of the family's primary philanthropies, particularly RBF" (Rockefeller, 2002, 339).

David remarked that a leftward change in JDR III's political views contributed to some of the disagreement. JDR III interviewed many young people in the early 1970s to seek and understand their point of view. John published what he learned in *The Second American Revolution* (Rockefeller, J.D. III, 1973).

JDR III expressed his conviction that the unrest on the streets of the United States in the 1960s and early 1970s signaled a turning point in history. He wrote that "instead of being overwhelmed by our problems, we must have faith that they can be resolved, that we can achieve a society in which humanistic values predominate" (Rockefeller, J.D. III, 1973, p. xiv).

JDR III found "two very broad sources of revolutionary change in our society. One is people, individually and in groups, who are concerned about justice and freedom and receiving a fair share of the fruits of our society. The other is impersonal and materialistic, stemming from economic growth, new knowledge and technological innovations, international rivalries" (Rockefeller, J.D. III, 1973, p. 5).

JDR III called the revolution a humanistic revolution since it arose from the first source of change: "the wants and needs and aspirations of people" (Rockefeller, J.D. III, 1973, p. 6). He considered the change to be a revolution because it had the potential to result in far-reaching social change and possibly the replacement of one government with another. He further characterized the revolution as a humanistic revolution because it would "harness the forces of economic and technological change in the service of humanistic values" (Rockefeller, J.D. III, 1973, p. 6).

According to JDR III, the central meaning of the youth movement was "to achieve a person-centered society, instead of one built around materialism and large impersonal institutions which breed conformity rather than individuality and creativity" (Rockefeller, J.D. III, 1973, p. 28). He believed that the youth movement should be supported rather than suppressed by the older generation because he could identify some consequences of the youth movement that he considered positive accomplishments. He said that the youth movement had raised public awareness of issues, that youth were more interested in ideas and spiritualism rather than materialism, and that the roots of poverty are "to be found less in the failings of the poverty-stricken than in social imbalances and discrimination" (Rockefeller, J.D. III, 1973, pp. 31–32).

JDR III understood that the "steady growth of centralized government must ultimately result in a statist and bureaucratic pattern quite unlike anything we have known in the past. The only alternative is for government and other sectors of society to collaborate in leading us back to a balanced system, based on the bedrock of individual initiative" (Rockefeller, J.D. III, 1973, p. 108). He wanted the United States government to make it easier for lower levels of government and the private sector to help solve social issues.

David Rockefeller wrote that JDR III believed that "all wisdom reposed in the young and that the older generation, which had made such a mess of the world, should look to their children for guidance" (Rockefeller, 2002, p. 339). JDR III's study of the youth movement had "strengthened his instinctive sympathies for the underdog and the underclass" (Rockefeller, 2002, p. 339). Nelson, David's brother, recognized an apparent shift in JDR III's views on politics. Consequently, Nelson did not want JDR III in a position that could aid groups that would oppose Nelson's political endeavors.

In an autobiography entitled *Memoirs*, David Rockefeller said that "For more than a century ideological extremists at either end of the political spectrum have seized upon well-publicized incidents such as my encounter with Castro to attack the Rockefeller family for the inordinate influence they claim we wield over American political and economic institutions. Some even believe we are part of a secret cabal working against the best interests of the United States, characterizing my family and me as 'internationalists' and of conspiring with others around the world to build a more integrated global political and economic structure — one world, if you will. If that's the charge, I stand guilty, and I am proud of it" (Rockefeller, 2002, p. 405).

David believed that the United States should embrace its global responsibilities. He said that "the free flow of investment capital, goods, and people across borders will remain the fundamental factor in world economic growth and in the strengthening of democratic institutions everywhere" (Rockefeller, 2002, 406). Therefore, he concluded that "we must all be internationalists" (Rockefeller, 2002, p. 406).

David Rockefeller died in 2017.

11.5 The Rockefeller Foundation and the World Economic Forum

The Rockefeller Foundation did not end with the death of John D. Rockefeller's last grandson David. Today, the Rockefeller Foundation promotes the use of science, data, policy, and innovation to find solutions to global challenges associated with food, power, and economic mobility. It considers itself a science-driven philanthropy that strives to identify and advance ground-breaking solutions, concepts, and interactions that can have a significant positive impact on the well-being of humanity worldwide (WEF RF, 2023). One way to achieve this influence is by contributing to other multinational organizations. For example, the Rockefeller Foundation is a contributor to the World Economic Forum (WEF), a notable and especially influential multinational organization.

11.5.1 *Klaus Schwab and Industrial Revolutions*

The WEF was founded as the non-profit European Management Forum by Klaus Schwab on January 24, 1971 (WEF Mission, 2023). Schwab, with doctorates in engineering and economics, was a pioneer in developing the fourth industrial revolution.

The First Industrial Revolution (FIR) began with the invention of steam engines. It was based on the application of steam power to a range of technologies including steam-powered factories and machine production. Abundant coal was used to provide motive power to a national rail system where locomotives were driven by steam resulting from coal combustion.

The Second Industrial Revolution (SIR) began with the harnessing of electricity. The SIR applied scientific advances to mass production and manufacturing. Generated electricity was used to power new inventions such as the telephone, radio, and television. Inexpensive oil was used by vehicles with internal combustion engines to transport people and goods on national road systems.

The First and Second Industrial Revolutions were part of the carbon economy. They relied on combustion of coal, oil, and natural gas. The digital era began the Third Industrial Revolution (TIR). The TIR was characterized by scientific advances in electronics, computing, and information technology. The use of electricity as a carrier of energy and innovations in renewable energy during the TIR are supporting a transition to a zero-carbon emission economy.

A Fourth Industrial Revolution (4IR) emerged in the 21st century. In an essay first written by Klaus Schwab, the digitization of information that began in TIR was extended in 4IR using such technologies as "artificial intelligence, genome editing, augmented reality, robotics, and 3D printing" (BOE 4IR, 2023). The essay recognized that 4IR involved "systemic change across many sectors and aspects of human life" (BOE 4IR, 2023). The 4IR was capable of transforming society on a global scale, but it comes with risks. Among the risks are "cybersecurity threats, misinformation on a massive scale through digital media, potential unemployment, or increasing social and income inequality" (BOE 4IR, 2023). On the other hand, the 4IR "is an opportunity to unite global communities, to build sustainable economies, to adapt and modernize governance models, to reduce material and social inequalities, and to commit to values-based leadership of emerging technologies" (BOE 4IR, 2023).

Table 11.1 presents an approximate timeline for five industrial revolutions. The Fifth Industrial Revolution (5IR) focuses on synergistic collaboration of humans and machines rather than focus on 4IR competition and possibly replacement of humans by machines. Furthermore, the 4IR

Table 11.1 Approximate Timeline for Industrial Revolutions (Noble *et al.*, 2022)

Industrial Revolution	Approximate Timeline
First (FIR)	ca. 1750–1850
Second (SIR)	ca. 1850–1930
Third (TIR)	ca. 1930–2000
Fourth (4IR)	ca. 2000–2020
Fifth (5IR)	ca. 2020–present

does not have an environmental emphasis, while the 5IR embraces a concern for humanity and the planet. The environmental emphasis of the 5IR includes a focus on sustainable and renewable resources (Noble *et al.*, 2022).

11.5.2 *The World Economic Forum*

The WEF is an international non-governmental and lobbying organization that supports cooperation between the public and private sectors. It is headquartered in Geneva, Switzerland. The institutional culture of the WEF is based on the stakeholder theory, which asserts that organizations should be accountable to all segments of society. WEF strives to promote entrepreneurship in the service of the general welfare on a global scale while maintaining the most rigorous principles of governance. It seeks to solve industrial, regional, and global issues by engaging leading figures from different walks of life who have the personal drive and organizational influence to achieve positive change (WEF Mission, 2023).

One of the WEF projects that is supported by the Rockefeller Foundation is the Commons Project, a non-profit public trust that has the purpose of developing digital services for the general good. The CommonHealth Android platform was developed under the auspices of the Commons Project. The platform lets users gather, manage, and share personal health information with trusted apps, services, and organizations (WEF Commons Project, 2023).

Another WEF project that is supported by the Rockefeller Foundation is GAEA (Giving to Amplify Earth Action). According to a WEF press release, GAEA "will leverage philanthropic capital to help generate the

$3 trillion needed each year from public and private sources to tackle climate change and nature loss" (WEF GAEA, 2023).

The GAEA initiative is supported by more than 45 major philanthropic, public, and private sector partners including the Rockefeller Foundation. The WEF press release reported that "Greater philanthropic funding for climate and nature will support, not detract from, existing social priorities. As recently noted by Rajiv Shah, President, The Rockefeller Foundation: 'Climate change poses a singular threat to humanity… we must directly confront climate change, even as we redouble efforts in our traditional program areas: health, power, food, and equity'" (WEF GAEA, 2023).

Chapter 12

Weaponizing National Governments

Globalists can use the bureaucracies of centralized national governments to seize control of sovereign states. Many authors have considered how bureaucracies and their sponsors have sought to control otherwise free states. In this chapter, we describe some notable examples of this mode of weaponization.

12.1 Eisenhower and the Military-Industrial Complex

On January 17, 1961, Dwight D. Eisenhower addressed the country in his final address as President of the United States (Eisenhower, 1961). Eisenhower was a retired five-star general. During World War II, Eisenhower was the military commander that led the allies to victory in Europe. In his farewell address following his second term as president, he cautioned that a military-industrial complex was emerging and could be a threat to democracy.

Eisenhower began his farewell address by saying that the history of the United States was an "adventure in free government." He said that "our basic purposes have been to keep the peace; to foster progress in human achievement; and to enhance liberty, dignity and integrity among people and among nations" (Eisenhower, 1961). He recognized that "A vital element in keeping the peace is our military establishment. Our arms must be mighty, ready for instant action, so that no potential aggressor may be tempted to risk his own destruction."

In a brief review of the history of armaments, Eisenhower pointed out that the United States did not maintain a permanent armaments industry until the world conflicts of the 20th century. It was now clear to him that "we can no longer risk emergency improvisation of national defense; we have been compelled to create a permanent armaments industry of vast proportions."

Eisenhower observed that "this conjunction of an immense military establishment and a large arms industry is new in the American experience." He introduced the term military-industrial complex to describe this conjunction of military and industrial strength. He then warned that "In the councils of government, we must guard against the acquisition of unwarranted influence, whether sought or unsought, by the military-industrial complex. The potential for the disastrous rise of misplaced power exists and will persist."

Eisenhower considered the infrastructure that developed sophisticated weaponry during the first half of the 20th century a technological revolution. The period saw the development of such weapons as radar, sonar, aircraft, aircraft carriers, rocketry, and nuclear weapons. He said that "research has become central; it also becomes more formalized, complex, and costly. A steadily increasing share is conducted for, by, or at the direction of, the Federal government." Furthermore, "The prospect of domination of the nation's scholars by Federal employment, project applications, and the power of money is ever present and is gravely to be regarded." This led to the concern that "we must also be alert… that public policy could itself become the captive of a scientific technological elite."

Referring to the conjunction of military and industrial systems, Eisenhower said "We must never let the weight of this combination endanger our liberties or democratic processes … Only an alert and knowledgeable citizenry can compel the proper meshing of the huge industrial and military machinery of defense with our peaceful methods and goals, so that security and liberty may prosper together."

Eisenhower expressed his vision for the future: "we — you and I, and our government — must avoid the impulse to live only for today … We cannot mortgage the material assets of our grandchildren without risking the loss also of their political and spiritual heritage." He hoped that "this world of ours … must avoid becoming a community of dreadful fear and hate, and be, instead, a proud confederation of mutual trust and respect."

Point to Ponder: Have Eisenhower's Hopes Been Realized?

Eisenhower's hope that the nations of the world would establish a "proud confederation of mutual trust and respect" is similar to hopes expressed by founders of multinational organizations such as the League of Nations and the United Nations. Events since Eisenhower's 1961 farewell address seem to support the view that the world is trending toward a collection of fearful and hostile states. The military-industrial complex, for example, appears to be one element of a deep state that is undermining freedom and democratic values, as we explain in the following sections.

12.2 Political Oligarchies

An oligarchy is a form of government in which the majority of the population is ruled by a small group of people known as oligarchs. The ruling class consists of people with the authority to manage public resources to the wealthy with the power to influence government decisions. Two types of political oligarchies are considered in this section.

The first type is the plutocracy model based on *Tragedy and Hope — A History of the World in Our Time* by Carroll Quigley, a Harvard historian (Quigley, 1966). Quigley described how decisions made by political leaders could be influenced by wealthy people.

The second type is the bureaucratic ruling class model based on *The Bureaucratization of the World* by Bruno Rizzi (1939, French edition). Two variations of the bureaucratic ruling class model are also considered: the managerial ruling class model based on *The Managerial Revolution* by James Burnham (Burnham, 1941), and the Deep State ruling class model (e.g., Lofgren, 2014; Chaffetz, 2018, 2023; Patel, 2023).

Rizzi and Burnham noted that public and private sector institutions tend to be governed by unelected bureaucrats with minimal accountability. In some cases, an elected official might be accepted into the ruling class if the elected official supported the ruling class.

The ruling class models outlined here help explain why a change in government from one political party to another in modern democracies does not always result in significant government policy changes. One reason for government inertia is the recognition that career bureaucrats can choose to support changes introduced by a newly installed government, or they can choose to impede reforms.

12.3 Quigley's Plutocracy Model

Historian Carroll Quigley was a professor at Princeton, Harvard, and the Georgetown University Foreign Service School. He wrote the text *The Evolution of Civilizations: An Introduction to Historical Analysis* (Quigley, 1961). In his 1966 book *Tragedy and Hope — A History of the World in Our Time* (Quigley, 1966), Quigley concentrated on the influence of wealth on politics from the middle of the 19th century until the middle of the 20th century.

The period from the mid-19th to the mid-20th century encompassed the American Civil War, World Wars I and II, and the Great Depression. Quigley believed that the barbarism of this period was a result of "materialism, selfishness, false values, hypocrisy and secret values" (Quigley, 1966, p. 1310). He wrote that humanity could create a better world based on such Western qualities as "generosity, compassion, cooperation, rationality, and foresight, and finding an increased role in human life for love, spirituality, charity, and self-discipline" (Quigley, 1966, p. 1311).

W. Cleon Skousen wrote a review of Quigley's 1300+ page book *Tragedy and Hope*. Skousen's review was entitled *The Naked Capitalist* (Skousen, 1970). He said that Quigley exposed the political impact of people who finance and significantly influence the political system. Another book entitled *None Dare Call It Conspiracy* (Allen and Abraham, 1971) also expressed concern about the influence of private interests in politics. It is apparent that Quigley's ideas about the significance of money and mass media in democratic systems should be considered.

Quigley spent 20 years studying the operation of a network of wealthy individuals and was allowed to "examine its papers and secret records" for 2 years in the early 1960's (Quigley, 1966, p. 950). Quigley's thesis was that a group of people called the network was using money to manipulate human affairs around the world. The network wanted to remain unpublicized, but Quigley argued that "its role in history is significant enough to be known" (Quigley, 1966, p. 950).

12.3.1 *John Ruskin and the Network*

The network studied by Carroll Quigley was formed in the latter half of the 19th century. Quigley wrote that the network was originally inspired by the ideas of John Ruskin (1819–1900). Ruskin graduated from Oxford University and was appointed Slade Professor of Art at Oxford in 1870

(Quigley, 1966, p. 130). He believed that the state should be governed by an elite ruling class, even if it only consisted of one person: "My continual aim has been to show the eternal superiority of some men to others, sometimes even of one man" (Clark, 1964, p. 267).

The influence of the network was extended by enlisting followers using education and purchasing support while other people fought for freedom and independence. Ruskin scorned freedom by comparing it to the freedom of the house fly: "I believe we can nowhere find a better type of a perfectly free creature than in the common house fly... There is no courtesy in him; he does not care whether it is king or clown he teases; and in every pause of his resolute observation, there is one and the same expression of perfect egotism, perfect independence and self-confidence, and conviction of the world's having been made for flies... You cannot terrify him, nor govern him, nor persuade him, nor convince him. He has his own positive opinion on all matters..." (Clark, 1964, p. 302).

In his book about Ruskin, Clark said that "the authoritarian element in Ruskin's ideal policy ... was derived directly from the source book of all dictatorships, Plato's *Republic*. He read Plato almost every day..." Furthermore, Clark wrote, "Plato wanted a ruling class with a powerful army to keep it in power and a society completely subordinate to the monolithic authority of the rulers" (Clark, 1964, p. 269).

Marriage and families would be prohibited in Plato's ideal society. All women would be able to claim all men and vice versa. The state would regulate selective breeding. The state would destroy inferior or disabled children and raise the remaining children once they were weaned. No distinction would be made between the roles played by men and women; both sexes would engage in labor and combat.

Plato's ideal society would consist of a hierarchy of three classes: the ruling class, the military class, and the working class. The ruling class would live in communal families, share property, and apply their intellectual energies to determining what was best for the lower classes.

When Ruskin presented his ideas at Oxford University, the pinnacle of English education in the 19th century, England was still a colonial power. He viewed Oxford students as members of the privileged, ruling class. Ruskin believed that the English ruling class should teach their values and traditions to the lower classes throughout the world, including England. Ruskin wanted to preserve the status of the ruling class by convincing the much more populous lower classes that they should embrace the values and traditions of the ruling class. Cecil John Rhodes

(1853–1902), one of Ruskin's first students, became a practitioner of Ruskin's views.

12.3.2 Cecil Rhodes

British-born Cecil Rhodes only spent one term at Oxford in 1873 before leaving for South Africa. While at Oxford, Rhodes attended Ruskin's inaugural lecture (NWE-Rhodes, 2018). Ruskin's lecture reinforced Rhodes' support for British imperialism. In 1876, Rhodes returned to Oxford for a second term. His time at Oxford influenced the development of his desire to expand British dominance throughout the world. This vision needed money.

Rhodes joined his brother Herbert in the diamond fields of Kimberley, South Africa in 1871 when he was 18 years old. Cecil oversaw work on Herbert's claim and supported Herbert's commercial speculation. Cecil's activities helped him become a millionaire by the time he was 20. He traveled to Oxford in 1873 after entrusting care of the Kimberley diamond fields to his brother Herbert and his associate Charles D. Rudd.

An economic depression in the diamond fields occurred in 1874 and 1875. The depression gave Cecil Rhodes an opportunity to increase his participation in the de Beers mine. Rhodes and Rudd started the De Beers Mining Company in 1880. NM Rothschild & Sons in London funded De Beers. NM Rothschild was Nathaniel Meyer Rothschild (1840–1915), a banker and politician who became the first Baron Rothschild.

Rhodes supported South African and English political allies with his fortune. From 1890 to 1896, he served as Prime Minister of Cape Colony. His decisions aided in the extension of British imperial policies in South Africa. One of his goals was to secure a strip of British territory from the Cape of Good Hope to Egypt. A railroad from one end of Africa to the other would be built on the strip. His adventurism provoked conflict with the Transvaal Boer Republic. Rhodes resigned as Prime Minister after an unsuccessful attack on the Transvaal in 1895.

Rhodes was able to acquire mineral concessions using his wealth and connections with the British government in Africa. As a result of his business acumen and government connections, he amassed a sizable fortune by the time his first will was written in 1877. Rhodes believed that the British were "the finest race in the world and that the more of the world we inhabit the better it is for the human race" (Flint, 1974). He wanted to use his fortune to bring the entire world under British dominion by founding a secret society (NWE-Rhodes, 2018):

> "To and for the establishment, promotion and development of a Secret Society, the true aim and object whereof shall be for the extension of British rule throughout the world, the perfecting of a system of emigration from the United Kingdom, and of colonisation by British subjects of all lands where the means of livelihood are attainable by energy, labour and enterprise, and especially the occupation by British settlers of the entire Continent of Africa, the Holy Land, the Valley of the Euphrates, the Islands of Cyprus and Candia, the whole of South America, the Islands of the Pacific not heretofore possessed by Great Britain, the whole of the Malay Archipelago, the seaboard of China and Japan, the ultimate recovery of the United States of America as an integral part of the British Empire, the inauguration of a system of Colonial representation in the Imperial Parliament which may tend to weld together the disjointed members of the Empire and, finally, the foundation of so great a Power as to render wars impossible, and promote the best interests of humanity."

Rhodes provided funds in his last will and testament for scholarships to Oxford University that are now known as Rhodes Scholarships. The scholarships were available to qualified students from territories under British rule, formerly under British rule, or Germany.

Cecil Rhodes died as one of the richest men in the world in 1902.

12.3.3 *Cecil Rhodes and the Secret Society*

Cecil Rhodes used his wealth to fund the mission of the Secret Society, which was created to support British rule over a New World Order. According to Quigley, Rhodes formed his Secret Society in 1891 at a London meeting with journalist William T. Stead, and Reginald Baliol Brett, who was also known as Lord Esher (Quigley, 1981, Chapter 1). Quigley described Brett as "friend and confidant of Queen Victoria, and later to be the most influential adviser of King Edward VII and King George V" (Quigley, 1981, p. 3).

The organization plan of the Rhodes Secret Society developed by Rhodes, Stead, and Brett included an inner circle called The Society of the

Elect, and an outer circle called The Association of Helpers. Rhodes led the Secret Society and a "Junta of Three" that included Stead, Brett and Alfred Milner (1854–1925). Milner, an associate of Baron Rothschild, helped found the Secret Society.

From 1891 until his passing in 1902, Rhodes served as leader of the Secret Society. Stead was the most influential member of the Secret Society during this period. After Rhodes' passing, Milner took over leadership of the Secret Society. Its goals remained the same under Milner's leadership, but not its culture. Quigley recognized the cultural change by referring to the organization as Rhodes Secret Society before 1901 and the Milner group thereafter (Quigley, 1981, p. 4).

According to Quigley (1966, p. 132): "As governor-general and high commissioner of South Africa in the period 1897–1905, Milner recruited a group of young men, chiefly from Oxford and from Toynbee Hall, to assist him in organizing his administration. Through his influence these men were able to win influential posts in government and international finance and became the dominant influence in British imperial and foreign affairs up to 1939. Under Milner in South Africa they were known as Milner's Kindergarten until 1910. In 1909–1913 they organized semisecret groups, known as Round Table Groups, in the chief British dependencies and the United States. These still function in eight countries. They kept in touch with each other by personal correspondence and frequent visits, and through an influential quarterly magazine, The Round Table, founded in 1910 and largely supported by Sir Abe Bailey's money. In 1919 they founded the Royal Institute of International Affairs (Chatham House) for which the chief financial supporters were Sir Abe Bailey and the Astor family (owners of *The Times*). Similar Institutes of International Affairs were established in the chief British dominions and in the United States (where it is known as the Council on Foreign Relations) in the period 1919–1927. After 1925 a somewhat similar structure of organizations, known as the Institute of Pacific Relations, was set up in 12 countries holding territory in the Pacific area, the units in each British Dominion existing on an interlocking basis with the Round Table Group and the Royal Institute of International Affairs in the same country. In Canada the nucleus of this group consisted of Milner's undergraduate friends at Oxford (such as Arthur Glazebrook and George Parkin), while in South Africa and India the nucleus was made up of former members of Milner's Kindergarten."

A group of scholars known as The Inquiry was assembled by Edward M. House (Gavrilis, 2021, Chapter 1) to help end World War I. The group was composed of experts on "the political, economic, and social conditions in war-ravaged Europe, especially the status of the territories of the losing side" (Gavrilis, 2021, pp. 2–3). Members of the group were part of the American delegation to the Paris Peace Conference (see Chapter 5 for more discussion).

The slow progress of negotiations in Paris during the early months of 1919 frustrated members of the Inquiry. They knew that "The American public was eager to put the ware behind it" (Gavrilis, 2021, pp. 3 4). Concerned about the direction of the conference, "the Americans who made up the Inquiry and a few diplomats they had befriended along with their like-minded British counterparts committed to stay in touch" (Gavrilis, 2021, p. 4). This enlarged group of British and American delegates met in May 1919 at the Hotel Majestic in Paris, a few weeks before the Treaty of Versailles was signed on June 28, 1919.

Lionel Curtis, a British diplomat, advocated for the creation of an institute for the study of international affairs. Curtis wanted the institute "to foster mutual understanding between nations and… to propose solutions to the biggest challenges facing the world" (Chatham House, 2023). The British and American delegates "proposed forming an Anglo-American Institute of International Affairs to sustain their work and suggest policies to guide the post-war order" (Gavrilis, 2021, p. 4).

The British contingent went home after the Paris Peace Conference and founded the British Institute of International Affairs in London in July 1920. The British Institute of International Affairs is now known as the Royal Institute of International Affairs, or Chatham House. The American contingent returned to America to find a climate of isolationism that was incompatible with their understanding of the ideals of the League of Nations. The work of the American contingent faltered.

Interest in an American institute of international affairs was revived by the launch of Chatham House; the creation of an American Institute of International Affairs by a New York woman who was not part of the American contingent in Paris; and the disbanding of a little-known club called the Council on Foreign Relations in early 1921. Members of the now defunct Council on Foreign Relations were "concerned that any turn away from internationalism would hurt postwar business" (Gavrilis, 2021, p. 8). The group was led by Elihu Root, a Nobel laureate and former Secretary of State during the Theodore Roosevelt administration.

According to Gavrilis, "Representatives of the Inquiry quickly reassembled and for five months discussed a merger with several members of the defunct New York club. They decided to form a new organization and to restrict membership to US citizens" (Gavrilis, 2021, p. 8). William H. Shepardson, an aide to Edward M. House in Paris and secretary of the commission that drafted the Covenant of the League of Nations, was asked to inform "the British counterparts of the Inquiry that it would no longer be possible to have a joint Anglo-American institute. He was relieved to learn on arriving in London that the British had made the same decision" (Gavrilis, 2021, p. 8). To avoid the appearance of any linkage with the British Institute of International Affairs, the "proposed American Institute of International Affairs was instead called the Council on Foreign Relations" (Gavrilis, 2021, p. 8). The Council on Foreign Relations was registered in New York on July 29, 1921.

The Paris Peace Conference that ended World War I also created The League of Nations. The League of Nations was founded to maintain world peace on January 10, 1920. As an international organization, the League of Nations was a forerunner of the United Nations. The Council on Foreign Relations was the United States affiliate of the Round Table with expertise in foreign affairs and international relations.

The Council on Foreign Relations participated in the 1945 United Nations Conference on International Organization in San Francisco. The Charter of the United Nations was signed at the conclusion of the conference.

12.4 The Bureaucratic Ruling Class Model

Communist intellectual Bruno Rizzi contended in *The Bureaucratization of the World* (Rizzi, 1939) that the Soviet Union under Joseph Stalin was not a Communist system but rather a new form of governing system he named "bureaucratic collectivism." Rizzi said the Soviets had established a new governing class made up of political and economic bureaucrats in place of the capitalist and aristocratic ruling classes. The bureaucrats were able to retain power by establishing a police state. Rizzi concluded that the Soviet Union was imposing a new system of exploitation "on the entire world" to defend itself (Rizzi, 1939, Section VIII). To Rizzi, "the Stalinist regime is intermediate, it throws aside outdated capitalism, but it does not rule out socialism for the future. It is a new social form, based on class property and class exploitation."

Bolshevik theoretician Leon Trotsky analyzed Rizzi's "Theory of Bureaucratic Collectivism" in *The USSR in War* (Trotsky, 1939). Trotsky referred to Bruno Rizzi as Bruno R. He did not agree with Rizzi that the Stalinist bureaucracy should be considered a new class. Trotsky wrote that the Stalinist regime was either "an abhorrent relapse in the process of transforming bourgeois society into a socialist society, or the Stalinist regime is the first stage of a new exploiting society." If the latter option is correct, Trotsky concluded that "the bureaucracy will become a new exploiting class" (Trotsky, 1939, p. 6)

James Burnham questioned Trotsky's assessment of Rizzi's bureaucratic collectivism (Burnham, 1941). Instead, he suggested that a social revolution was creating a new ruling class. He identified three elements of the social revolution (Burnham, 1941, p. 5):

1. A drastic change occurs in the most important social (economic and political) institutions.
2. Concurrent with change in social institutions is a shift in the cultural institutions and dominant beliefs about the role of humanity in our world and the universe.
3. A change in the group of people who control the greater part of power and privilege in society.

Burnham postulated that he was living through a time of social revolution: society was transitioning from one kind to another (Burnham, 1941, p. 7). Burnham developed his theory of the managerial revolution based on these observations. His theory was designed to explain the societal transition and anticipate the kind of society that would emerge. He thought the transition would be from a capitalist or bourgeois society to a society he called managerial (Burnham, 1941, p. 71).

According to Burnham, an emerging managerial class would take power following an ongoing global struggle for social dominance during the transition. The managerial class would achieve social dominance by using the state to control the means of production. It would become the ruling class (Burnham, 1941, p. 72). The people who owned the means of production under capitalism would cede control to the managerial class (Burnham, 1941, Chapter VII).

Trotsky refuted the theories advocated by Rizzi, Burnham, and their allies in his book *In Defense of Marxism* (Trotsky, 1942). Trotsky thought Burnham was anti-dialectic because he did not use the dialectical method of reconciling two opposing points of view. In the dialectical method, the opposing viewpoints would be considered a thesis and its antithesis. The synthesis would then reconcile the opposing viewpoints. Trotsky concluded that Burnham's reasoning and results were incorrect because Burnham was not properly applying dialectics.

The way they viewed Stalin's totalitarian regime marked the primary distinction between Trotsky and Burnham. Trotsky saw Stalin's regime as a government endeavoring to change from a bourgeoisie society to an egalitarian one. Burnham saw Stalin's regime as a new form of oppression. Author George Orwell fictionalized the viewpoint advocated by Burnham and his allies in *Animal Farm* (Orwell, 1945).

12.4.1 *Managerialism*

Burnham thought that capitalism was being replaced by a managerial revolution called managerialism (Burnham, 1941). Author George Orwell (actual name Eric Blair, 1903–1950) succinctly summarized Burnham's view (Orwell, 1946):

> "Capitalism is disappearing, but Socialism is not replacing it. What is now arising is a new kind of planned, centralized society which will be neither capitalist nor, in any accepted sense of the word, democratic. The rulers of this new society will be the people who effectively control the means of production: that is, business executives, technicians, bureaucrats, and soldiers, lumped together by Burnham, under the name of 'managers.' These people will eliminate the old capitalist class, crush the working class, and so organize society that all power and economic privilege remain in their own hands. Private property rights will be abolished, but common ownership will not be established. The new 'managerial' societies will not consist of a patchwork of small, independent states, but of great super-states grouped round the main industrial centers in Europe, Asia, and America. These super-states will fight among themselves for possession of the remaining uncaptured portions of the earth, but will probably be unable to conquer one another completely. Internally, each society will be hierarchical, with an aristocracy of talent at the top and a mass of semi-slaves at the bottom."

A century after the Fabian Society was established in 1884, Orwell published his classic dystopian novel *1984* (Orwell, 1949) which many view as a warning about life in a society founded on Fabian Socialism. Orwell was a socialist who recognized that not all socialists had the same point of view. He had reservations about socialists that were members of the bourgeois rather than the working class.

Orwell expressed his political views in his 1937 book *The Road to Wigan Pier* (Orwell, 1937, pp. 161–162):

> "Sometimes I look at a Socialist — the intellectual, tract-writing type of Socialist, with his pullover, his fuzzy hair, and his Marxian quotation and wonder what the devil his motive is. It is often difficult to believe that it is a love of anybody, especially of the working class, from whom he is of all people the furthest removed. The underlying motive of many Socialists, I believe, is simply a hypertrophied sense of order. The present state of affairs offends them not because it causes misery, still less because it makes freedom impossible, but because it is untidy; what they desire, basically, is to reduce the world to something resembling a chessboard. Take the plays of a lifelong Socialist like (George Bernard) Shaw. How much understanding or even awareness of working-class life do they display?"

Orwell concluded: "The truth is that, to many people calling themselves Socialists, revolution does not mean a movement of the masses with which they hope to associate themselves; it means a set of reforms which 'we', the clever ones, are going to impose upon 'them', the Lower Orders" (Orwell, 1937, p. 162).

12.4.2 *The Deep State*

Political authority in democracies typically rests with elected officials. The Deep State refers to individuals and organizations that exercise political power independent of, and sometimes in opposition to, elected officials. Dwight D. Eisenhower expressed concern about the military-industrial complex, a type of Deep State, that emerged during and after World War II. More recently, the 9-11-2001 terrorist attacks on the United States led to a security and intelligence operation that seems to be unaccountable to elected officials and the civilian legal system. Polls have shown that there is growing public concern that an American Deep State exists and may be expanding.

Mike Lofgren analyzed the Deep State in the United States based on 28 years' experience as a Congressional staff member, top security clearance, and national security expertise. According to Lofgren (Lofgren, 2014), "All complex societies have an establishment, a social network committed to its own enrichment and perpetuation."

The American Deep State is unique because of its scope, financial resources, and global reach. Lofgren said that the American Deep State "is a hybrid of national security and law enforcement agencies: the Department of Defense, the Department of State, the Department of Homeland Security, the Central Intelligence Agency, and the Justice Department. I also include the Department of the Treasury because of its jurisdiction over financial flows, its enforcement of international sanctions and its organic symbiosis with Wall Street."

Lofgren believed that the Executive Office of the Obama Administration's National Security Council coordinated these agencies. He said that some key areas of the judiciary are members of the Deep State and cited the Foreign Intelligence Surveillance Court (the FISA court) as an example. Another component of the Deep State included some congressional leaders and members of the defense and intelligence committees.

Jason Chaffetz, a former Utah Congressman, presented a view of the Deep State similar to Lofgren's. Chaffetz based his view of the Deep State on his experience in Congress during the Obama Administration (Chaffetz, 2018). He argued that America's federal bureaucracy has been "politicized and weaponized on behalf of one political party" (Chaffetz, 2023, p. xii). In a later book entitled *The Puppeteers*, Chaffetz extended his discussion of the exercise of political power by people and organizations that influence a deep state consisting of the federal bureaucracy and its private-sector loyalists (Chaffetz, 2023).

Like Chaffetz, Kash Pramod Patel presented his view of the Deep State, including participants and methods used by an entrenched government bureaucracy (Patel, 2023). Patel served as former chief of staff to the Department of Defense, deputy assistant to the president and senior director for counterterrorism on the National Security Council (NSC) in the Trump Administration. According to Patel, "the Deep State is the politicization of core American institutions and the federal government apparatus by a significant number of high-level cultural leaders and officials who, acting through networks, disregard objectivity, weaponize the law, spread disinformation, spurn fairness, or even violate their oaths of office for political and personal gain, all at the expense of equal justice and American national security" (Patel, 2023, p. 12).

Kash Patel was not the first to criticize the behavior of leaders in the modern ruling class. For example, Tucker Carlson expressed concern that the behavior of the ruling class was leading the United States to the brink of revolution (Carlson, 2018). As another example, Peter Schweizer examined the behavior of eight public figures to show how power could "be twisted to serve the ends of those who seek more of it" (Schweizer, 2020, p. 250). A few years later, Schweizer wrote in *Red Handed* that "We have exposed how elites, from Washington to Wall Street, from Silicon Valley to academe, have been coopted and are helping the regime (in China) while in some cases even bolstering China's military and intelligence complex" (Schweizer, 2022, p. 236).

If the above views of the Deep State are correct, or nearly so, it is reasonable to expect the Deep State to adapt when there is a change in leadership in the Executive Branch. Assuming the Deep State wanted to maintain and grow its power base, the Deep State would tend to resist a president that wanted to exercise more civilian control over the bureaucracy. By contrast, the Deep State would tend to support a president that wanted to expand federal government spending.

Another aspect of the Deep State that can appear in a multi-party, political system is the uniparty. The uniparty consists of influential individuals from more than one party in a multi-party system. The uniparty is essentially a form of Deep State when the uniparty can exercise political power regardless of which political party or candidate is elected. An examination of the size of the public sector as a fraction of gross domestic product (GDP) suggests that the United States after World War II can be considered an example of a uniparty system. Figure 12.1 compares United States government outlays from 1930 to 2022 and the political party controlling the presidency. The letter "D" denotes Democrat, and the letter "R" denotes Republican. It appears that the size of government tends to grow regardless of which political party is in control of the Executive Branch. A notable exception occurred during the Clinton Administration (D) in the 1990s when the House of Representatives was led by Speaker of the House Newt Gingrich (R).

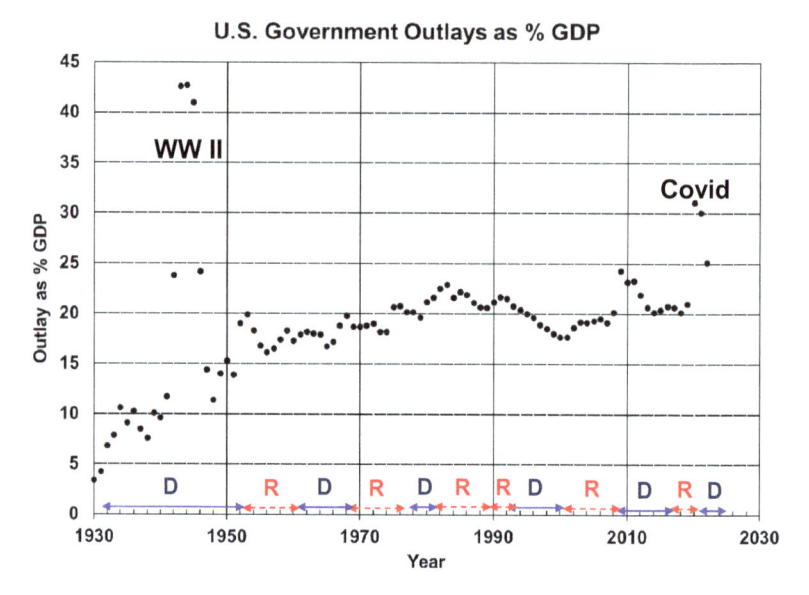

Figure 12.1 US Government Outlays as a Percent of GDP

Chapter 13
Weaponizing Human Conflicts

Human conflicts can be used by globalists to provide opportunities for seizing power from sovereign states. Several examples of human conflicts are presented here. In each case, the conflict raised concerns about the stability of a sovereign state.

13.1 The Leftist Attack on Christian America

Author David Horowitz argued that a war is being waged against American Christians by the political left. Horowitz wrote that the war "is not merely a war against an embattled religion. It is a war against an imperiled nation — a war against this nation and its founding principles: the equality of individuals and individual freedom. For these principles are indisputably Christian in origin. They are under siege because they are insurmountable obstacles to radicals' totalitarian ambition to create a new world in their image" (Horowitz, 2018, p. 22). In this section, we highlight key elements of Horowitz's argument and identify a few of the tactics that are being used to fundamentally transform the United States.

13.1.1 *The Mayflower Compact*

David Horowitz identified religious liberty as the founding principle of America (Horowitz, 2018). He said that America's founders were primarily Protestants escaping oppression by state-sanctioned religions. He illustrated this point by referring to the arrival of the Pilgrims at Plymouth

Rock in 1620. The Pilgrims were fleeing persecution by the Church of England, which was founded in 1534.

The Pilgrims crossed the Atlantic on the Mayflower and landed at Cape Cod. According to the Mayflower Society, "In 1620, 'Virginia' extended far beyond its current boundaries and the Mayflower was originally meant to land at its 'northern Parts,' specifically the Hudson River. When the Mayflower attempted to sail around Cape Cod to reach the Hudson, contrary winds and dangerous shoals forced the ship to turn around and instead anchor in modern day Provincetown Harbor on November 11, 1620" (Mayflower Compact, 1620).

Before disembarking, most of the male passengers on the Mayflower signed a document which is now known as the Mayflower Compact. The Mayflower Compact was signed on November 11, 1620, and is shown in Table 13.1. It helped explain why the Pilgrims were willing to undertake the dangerous journey to America. William Bradford (1590–1657), governor of Plymouth Colony for three decades, wrote that the Mayflower Compact was signed before coming ashore and served as "the first foundation of their government in this place" (Mayflower Compact, 1620).

Horowitz considered the Mayflower Compact a social contract and called the government of Plymouth Colony "the first civil government

Table 13.1 The Mayflower Compact (Mayflower Compact, 1620)

"Having undertaken for the Glory of God, and Advancement of the Christian Faith, and the Honour of our King and Country, a Voyage to plant the first Colony in the northern Parts of Virginia; Do by these Presents, solemnly and mutually, in the Presence of God and one another, covenant and combine ourselves together into a civil Body Politick, for our better Ordering and Preservation, and Furtherance of the Ends aforesaid: And by Virtue hereof do enact, constitute, and frame, such just and equal Laws, Ordinances, Acts, Constitutions, and Officers, from time to time, as shall be thought most meet and convenient [for] the general Good of the Colony; unto which we promise all due Submission and Obedience."

"IN WITNESS whereof we have hereunto subscribed our names at Cape-Cod the eleventh of November, in the Reign of our Sovereign Lord King James, of England, France, and Ireland, the eighteenth, and of Scotland the fifty-fourth, Anno Domini; 1620."

established by a social contract. The contract embodied the idea that governmental authority derives from the consent of the governed and that all its citizens are entitled to equal treatment under the law. These ideas became the core principles of the future American nation" (Horowitz, 2018, p. 42).

13.1.2 *The Establishment Clause*

The colonists in the New World had reached an important conclusion by 1776. Their experiences with colonial governance had convinced them that theocratic rule was undesirable and anti-Christian. Theocratic rule was oppressive and did not permit dissent or personal freedom of conscience. If the colonists wanted religious freedom for everyone, they needed to create a secular republic. Horowitz viewed this progression of ideas as "the logical, if not inevitable, development of the Protestant Reformation" (Horowitz, 2018, p. 43).

Horowitz cited two core doctrines of Protestant Reformation to justify his claim. The first doctrine is "justification by faith." It says that God's grace is the only way people can be saved because we are naturally imperfect and sinful. The first doctrine supported the American belief that government must have a system of checks and balances to manage the nefarious inclinations of government officials.

The second doctrine is "the priesthood of all believers." It rejected the belief that an organization of people, such as the leadership of the Catholic Church, was needed to mediate a person's relationship with God. The second doctrine supported two key principles of the American Declaration of Independence: all men are created equal, and they are endowed by their Creator with certain unalienable rights that government did not have authority to refuse. Horowitz said that this proclamation "led to the abolition of slavery and the enfranchisement of women" (Horowitz, 2018, p. 46).

The First Amendment of the Bill of Rights says that "Congress shall make no law respecting an establishment of religion, or prohibiting the free exercise thereof; or abridging the freedom of speech, or of the press; or the right of the people peaceably to assemble, and to petition the Government for a redress of grievances" (Bill of Rights, 1789). This Amendment prohibits the establishment of a state religion and protects every American's right to express and practice their religious beliefs. Horowitz argued that "The framers of the Constitution wrote the First

Table 13.2 Voluntary Prayer of the New York Board of Regents (Horowitz, 2018, p. 49)

"Almighty God, we acknowledge our dependence upon Thee, and we beg Thy blessings upon us, our parents, our teachers, and our country. Amen."

Amendment to prevent the suppression of religious freedom that is now epidemic in American schools" (Horowitz, 2018, p. 49).

A group of five people filed a legal complaint against the New York Board of Regents because they claimed the Board wrote a voluntary prayer that infringed on their right to freedom of religion. The complete text of the prayer is shown in Table 13.2. This case became known as Engel v. Vitale and was considered by the United States Supreme Court in 1962. The resolution of the case depended on a letter written by Thomas Jefferson to the Danbury Baptist Association of Connecticut on January 1, 1802.

In the letter, Jefferson said that "Believing with you that religion is a matter which lies solely between Man & his God, that he owes account to none other for his faith or his worship, that the legitimate powers of government reach actions only, & not opinions, I contemplate with sovereign reverence that act of the whole American people which declared that their legislature should 'make no law respecting an establishment of religion, or prohibiting the free exercise thereof,' thus building a wall of separation between Church & State" (Jefferson, 1802). The phrase "wall of separation between church and state" is now known as the Establishment Clause to denote separation of church and state.

Nine appointed justices considered the case and six agreed that the policy of the school system violated the Establishment Clause. In the process, they changed national law by creating a "constitutional right" that allowed a non-believing minority to restrict the freedom of the majority. Americans were no longer allowed to express their faith in public school.

The Establishment Clause has been embraced by secularists and the political left as a tool to marginalize and silence religious people. Schools risked violating Engel v. Vitale if they discussed the religious origins of American freedoms, such as the Mayflower Compact. Horowitz observed that "Censorship and the rewriting of history are the practices of totalitarian regimes" (Horowitz, 2018, p. 57).

13.1.3 *The Radical Left*

David Horowitz said that contemporary political divisions in America can be attributed to the radical left's resistance to the anti-Communist Vietnam war in the 1960s (Horowitz, 2018, p. 148). The radical left rioted outside of the 1968 Democratic Convention in Chicago. Then-President Lyndon Baines Johnson became President after the assassination of John F. Kennedy in 1963. Johnson chose not to run for reelection as president because there was little political support for his escalation of the Vietnam war. The radical left demonstrated against the 1968 Democrat nominee for President, then-Vice President Hubert Humphrey.

Humphrey was defeated by Richard Nixon in the 1968 general election in part because of Humphrey's support for America's involvement in Vietnam. Many members of the radical left became faculty in academia, members of the media, and took over leading positions in the Democrat Party following Humphrey's defeat. In addition to changing the structure and ideology of the Democrat Party, the radical left began changing the culture through academia, the media, and government.

The radical left established congressional caucuses based on identifying characteristics such as race, ethnicity, gender, and ideology. The caucuses served as platforms for fomenting dissent between identity groups and promoting progressive ideology.

It is worth noting here that David Horowitz was born to parents Phil and Blanche Horowitz. His parents were long-standing members of the American Communist Party and supported Joseph Stalin until they learned of Stalin's crimes against the Soviet people. Horowitz said that he "eventually came to understand that (his) parents and friends referred to themselves as 'progressives' to hide their true faith, which was communism" (Horowitz, 2018, p. 33).

According to Horowitz, progressives view people as naturally good, but can commit bad deeds because of society. They believe that social justice can be achieved by eliminating inequalities and oppressive behaviors that have troubled humanity throughout history. Progressives want to save society. "In contrast to the progressive mission of saving 'society,'" said Horowitz, "the goal of Christian belief is saving individual souls. Christians see the imperfections and suffering of the world as the results of acts by individuals who have failed to do good or have chosen to do evil" (Horowitz, 2018, p. 38). Unlike progressives, conservatives tend to believe that human beings are endowed with free will, and injustice can flow from the interaction of human beings exercising their free will.

Progressives and conservative Christians have opposing points of view. Horowitz observed that progressives are social redeemers that "view the Christian concerns for the salvation of individual souls as… a cause of social oppression" (Horowitz, 2018, p. 38). Therefore, the influence of conservative Christians in society must be removed.

The radical left, which is predominately progressive, took over the Democrat Party and replaced traditional liberalism "with a leftism that views American society as a system of oppressive hierarchies based on race, gender, and sexual orientation" (Horowitz, 2018, p. 148). The radical left brought with them "a collectivist ideology, rooted in Marxism, that is opposed to the American ideas of individual rights, individual accountability, and individual equality" (Horowitz, 2018, p. 148). Former Reagan Administration official Mark Levin pointed out that many American Marxists refer to themselves with terms such as progressives, democratic socialists, social activists, or community activists to cloak their political beliefs (Levin, 2021, p. 2).

American Marxism morphed into Cultural Marxism when Marxists replaced the struggle between classes based on economics by a struggle between groups based on identity (Gonzalez and Gorka, 2022). Cultural Marxism pits one group against another using identity characteristics "such as race, sex, or national origin, that are inherited at birth and over which the individual has no control" (Gonzalez and Gorka, 2022, p. 3).

13.1.4 *Roe v. Wade*

A movement known as the New Left arose during the 1960s and 1970s in Europe and the United States (BOE New Left, 2023). The emergence of the New Left coincided with the battle to prohibit prayer in public schools. The New Left in the United States grew out of the activism of socialist students, such as the Students for a Democratic Society (SDS). In an authoritative history of the movement that led to the Roe v. Wade decision by the Supreme Court, David J. Garrow traced the origin of the Roe v. Wade decision to a few women at the University of Texas at Austin (Garrow, 1994). The women were active in the SDS.

The culture war over abortion led to the Roe v. Wade decision by the Supreme Court. Pro-choice advocates, typically on the political left, tended to view abortion as a birth control method, and that there was little difference between abortion and contraception. Pro-life advocates, typically on the political right, recognized that there was a significant difference between abortion and contraception: abortion included another party, the unborn child.

In 1973, the Supreme Court agreed with pro-choice advocates that abortion was a constitutional right embedded in the pregnant woman's constitutional right to privacy. According to David Horowitz, "The secular left had discovered an all-powerful instrument — the Supreme Court — with which it could impose its radical, anti-Christian agenda on an unwilling nation" (Horowitz, 2018, p. 83). From this perspective, the Roe v. Wade decision "divided the entire nation because it involved the sanctity of life itself" (Horowitz, 2018, p. 78).

The American social contract had relied on democratic persuasion based on the consent of the governed and free speech for more than two centuries. The Supreme Court undermined the social contract when it imposed its will on America through the Roe decision and other similar rulings.

When Ronald Reagan was elected president in 1980, he hoped to nominate Supreme Court justices who shared his view that justices would interpret law as it was originally intended rather than write new law. During his first term as President, Donald Trump was able to achieve what Reagan could not. Trump was able to appoint enough new justices to change the ideological composition of the Supreme Court. The new majority did not see the right to abortion in the constitution and overturned Roe v. Wade in 2022.

13.1.5 *What Kind of Country Are We?*

David Horowitz wrote that Democrat politician Kamala Harris "is a supporter of racial preferences, a practitioner of racial politics, and an advocate of open borders" (Horowitz, 2018, p. 156). She has exhibited these characteristics as Joe Biden's Vice President. Speaking at a Howard University commencement in 2017, then-Senator Kamala Harris told her audience that "we have a fight ahead. This is a fight to define what kind of country we are."

The Constitution defines America as a nation that was defined when it was established. America is a country that cherishes individual freedom and religious liberty. Progressives and the Democrat Party have been trying to change the identity of America. According to Horowitz, "the Democrats offer us a reversion to tribal loyalties and collectivist values" (Horowitz, 2018, p. 157). Horowitz concluded that "A nation divided by such fundamental ideas — individual freedom on one side and group identity on the other — cannot long endure" (Horowitz, 2018, p. 157).

Point to Ponder: Who was Manning Johnson?

David Horowitz wrote with passion in his 2018 view of the leftist war against Christian America. His view was reminiscent of a view expressed by Manning Johnson in 1958 (Johnson, 1958, Chapter 1). Manning was a communist who wanted to replace capitalism in America with communism: "to me," he said, "the end of capitalism would mark the beginning of an interminable period of plenty, peace, prosperity and universal comradeship. All racial and class differences and conflicts would end forever after the liquidation of the capitalists, their government and their supporters."

Manning, an African American, began working for the red (communist) movement when he joined a front organization called the American Negro Labor Congress. This organization was designed to "trap the naïve, unwary, unsuspecting and idealistic Negro."

Manning spent 10 years working for the "Red Conspiracy," a movement of conspirators that were exploiting legitimate grievances "to transform idealism into a cold and ruthless weapon against the capitalist system." He "saw communism in all its naked cruelty, ruthlessness and utter contempt of Christian attributes and passions."

Manning's support of communism led to a rapid climb to membership in the National Committee, the most important governing body of the Communist Party in America. As a leading Negro communist, he served as a member of the National Negro Commission, an important sub-committee of the National Committee. His work was sufficiently noteworthy to the Kremlin leadership in the Soviet Union that Manning became a candidate for the Political Bureau (Politburo) of the Soviet Communist Party.

Manning worked with Kremlin representative Gerhard Eisler, who was assigned as a political commissar in America. Eisler was dissatisfied with the Kremlin because he believed the Soviet Communist Party was failing to take advantage of progressive groups. Eisler wanted to provide firm communist direction to sympathetic progressive groups to facilitate their control of the masses. Manning concluded that "as a participant on the highest level of the communist conspiracy in America, I observed the cold, calculating, ruthless nature of red power politics and political warfare, stripped of its illusory propaganda and idealistic cover."

Manning abandoned communism and wrote *Color, Communism and Common Sense* in 1958. His experience with communism was like author George Orwell's experience with Fabian Socialism described in Chapter 12.

13.2 EU 2050 and the Russia–Ukraine War

A transnational energy project called EU2050 (2023) is a project that spans national and continental boundaries. It serves as an illustration of a global effort to carry out the transition to a more sustainable energy portfolio. In this section, we outline EU 2050 and show that a contemporary European war can negatively impact the EU 2050 energy transition.

13.2.1 *European Super Grid*

The electrical grids of Europe were badly damaged during World War II. An organization called the Union for the Coordination of Production and Transmission of Electricity (UCPTE) was established in 1951. The UCPTE became an association of transmission system operators, or TSOs, in 1999 when the European Union deregulated electricity markets. It became the Union for the Coordination of Transmission of Electricity (UCTE) and unified the electrical grids of the member states of the European Union. The UCTE electrical grid has been synchronized through the Gibraltar AC link with the electrical grid in the North African nations of Algeria, Morocco, and Tunisia. Today, the UCTE is part of the European Network of Transmission System Operators for Electricity (ENTSO-E).

ENTSO-E has the mission of ensuring "the security of the interconnected power system in all time frames at pan-European level and the optimal functioning and development of the European interconnected electricity markets, while enabling the integration of electricity generated from renewable energy sources and of emerging technologies" (ENTSO-E, 2022, p. 2).

The vision of ENTSO-E is to play "a central role in enabling Europe to become the first climate-neutral continent by 2050 by creating a system that is secure, sustainable and affordable, and that integrates the expected amount of renewable energy, thereby offering an essential contribution to the European Green Deal. This endeavor requires sector integration and close cooperation among all actors" (ENTSO-E, 2022, p. 2).

The mission and vision of ENTSO-E assume that electricity will be the primary energy carrier in a carbon-neutral economy. The goal of ENTSO-E is to complete a fully carbon-neutral economy by 2050 that can accommodate variability and uncertainty in demand, generation, and grid availability.

The ENTSO-E power system will consist of three primary characteristics:

1. It will rely on carbon-neutral energy sources such as wind, solar, and geothermal.
2. It will be adaptable to variations in generation and consumption.
3. It will include a power grid that connects generators, consumers, and other resources throughout Europe.

The existing electrical grid of all participating states will become part of a Super Grid that incorporates direct current (DC) interconnectors. It will be able to integrate electricity from renewable energy sources. For example, the Super Grid could transmit electricity generated by offshore wind farms in the North Sea, solar power plants in Spain and North Africa, and French nuclear fission reactors.

Table 13.3 highlights some of the advantages and disadvantages of the EU Super Grid. The impact of change in EU membership was shown when the British voted to exit the EU on June 23, 2016. The EU had to decide if it should work with Britain to continue British implementation of the Super Grid. On the other hand, Britain had to decide if it wanted to continue participating in the Super Grid.

The Russian invasion of Ukraine in February 2022 has shown that grid vulnerability and dependence on non-member states are significant issues. It also illustrates the importance of fossil fuels in modern warfare. In the following sections, we summarize the status of electric vehicles (EVs) and then compare the military value of EVs and fossil fuel-powered vehicles.

Table 13.3 Advantages and Disadvantages of the EU Super Grid

Advantages	Disadvantages
Enhance the EU electricity market	Make the EU dependent on energy from non-member states
Support highly skilled jobs	Will be vulnerable to attack
Create opportunities to market sustainable energy technology by participating companies	Require significant expenditures by participating states

13.2.2 Electric Vehicles

There are many types of EVs available today. A hybrid vehicle supplements a gasoline-powered internal combustion engine (ICE) with a battery-powered electric motor. Hybrid vehicles with regenerative braking can generate electricity for recharging their batteries when the brakes are applied.

A plug-in hybrid electric vehicle (PHEV) uses a gasoline-powered ICE to supplement its battery-powered electric motor. The PHEV battery can be charged by connecting it to an external power source known as an EV charger. The battery in a PHEV has more storage capacity than a hybrid battery. The PHEV operates like an EV as long as its battery is charged. The ICE takes over when the battery runs down. The operator of a PHEV must recharge the battery and refill the gas tank. The PHEV has more range than an EV because it has a gasoline-powered ICE.

An EV is solely powered by a battery-powered electric motor. The storage capacity of the battery determines the range and performance of the EV. The EV must be recharged. The relative merits of EVs compared to ICE vehicles, hybrid vehicles, and PHEVs are discussed elsewhere (US EPA EV Myths, 2023; Igini, 2023; Fanchi, 2024). It is sufficient here to note that EVs depend on access to electricity. An EV charging station provides electrical energy to the EV battery. EVs must have access to EV charging stations, or they will not operate when the battery is depleted.

The charging rate is not the same for all battery chargers. Battery chargers, such as the Level 3 (DCFC) charger and Tesla supercharger, can charge an EV battery in less than an hour. Other battery chargers, such as the Level 1 and Level 2 chargers, are less expensive to own but take several hours to charge a battery.

The availability of public EV charging stations is increasing in some countries. The technical requirements to use a battery charger depend on factors such as current and voltage. Consequently, different plugs or adaptors may be needed to use a battery charger, otherwise the battery charger could be incompatible with the battery in an EV. More details are available in the references (Weber, 2021; NRDC, 2022).

> **Point to Ponder: Can the Government Impose Acceptance of EVs?**
>
> The government can attempt to impose the acceptance of EVs by passing laws and imposing regulations. For example, the California Air Resources Board of the State of California has adopted a set of regulations known as the Advanced Clean Cars II regulations. The regulations require all new passenger cars, trucks and SUVs sold in California to be zero-emission vehicles by 2035 (CARB, 2023). It remains to be seen if such regulations will be successful because demand for EVs seems to be settling at approximately 7% market share (Reuters EV Sales, 2023).

13.2.3 *Military Vehicles and the Russia–Ukraine War*

Nations around the world depend primarily on vehicles with ICEs. ICE vehicles that move on the ground typically use combustible gasoline or diesel. Aircraft tend to use combustible kerosene-based fuel. Gasoline, diesel, and kerosene are obtained from petroleum. The development and adoption of EVs is intended to replace ICE vehicles in a sustainable energy future. This raises the question: would EVs be preferable to ICE vehicles in war? The Russian invasion of Ukraine in 2022 suggests that EVs may not be well suited for military operations if ICE vehicles are available.

The invasion of Ukraine by Russia began as a military operation in 2022. It grew into an ongoing large-scale ground war. North Atlantic Treaty Organization (NATO) member countries responded to Russia's invasion by imposing sanctions on Russia and providing over $100 billion in financial and military aid to Ukraine. The sanctions caused a shortage of energy in Europe by disrupting the flow of Russian oil and natural gas to Europe. Many European countries had to contend with energy shortages during the winter months due to a scarcity of critical energy resources.

We need to consider the Russian invasion of Ukraine in 2022 in the context of the EU effort to transition to renewable energy by 2050. EU 2050 is designed to transition from fossil fuels to renewable energy resources. The European reliance on Russian oil and natural gas has left European countries exposed to disruptions in the energy supply when sanctions are imposed. The resulting energy shortage in Europe was aggravated because many European nuclear fission and coal power plants were decommissioned as part of the EU 2050 transition.

Nations without dependable fossil fuel supplies are significantly disadvantaged during a military conflict. The existing infrastructure in most nations is designed to transport and store oil and gas. In addition, generation and transmission of electricity from renewable energy sources and nuclear fission power plants are vulnerable to disruption or destruction during war. Modern EVs are not ideal for military operations. They rely on the availability of electricity and have relatively long battery charging times. The range of EVs is usually less than the range of ICE vehicles.

Contemporary military operations depend on the use of fossil fuels for fueling military vehicles and powering military bases. The high energy density of fossil fuels has justified the use of fossil fuels and ICE military vehicles. Gasoline and diesel are typically used in ground-based transportation. Kerosene-based fuel is used in aircraft.

Kerosene is a flammable liquid derived from the refining of petroleum. Kerosene is useful in aircraft because it has a lower viscosity and lower volatility than gasoline. Consequently, kerosene can be used at higher elevations where lower temperatures can cause gasoline to freeze.

The combat capability of a nation's military is impeded when its military does not have enough energy to operate and has less mobility than its adversary. Furthermore, the electricity generation and transmission system that provides electricity to a military that relies on EVs is subject to targeting by its adversary. The Russian attack on power facilities and transmission lines has reminded us of the strategic importance of electricity in warfare.

13.2.4 *Pipelines and the Russia–Ukraine War*

The European effort to implement EU 2050, a multinational energy transition project, requires close coordination between many different countries. EU 2050 experienced unanticipated challenges following a sequence of incidents that started in 2021.

Joe Biden was inaugurated as President of the United States in January 2021. Following inauguration, the Biden Administration stopped the completion of the Keystone pipeline from Canada to the United States. It then lifted sanctions on a natural gas pipeline from Vyborg, Russia through the Baltic Sea to Lubmin, Germany called the Nord Stream 2 (Amaro, 2021). These decisions should be understood in a historical context.

In 2013, Ukrainian Crimea was annexed by Russia. Russia and Ukraine were members of the Soviet Union until the Soviet Union collapsed in the 1980s. Russia and Ukraine emerged as sovereign states. Russian natural gas was transported to several EU member states through a gas pipeline network in Ukraine. Europeans used gas for heat during the winter. Russia threatened to stop supplying gas to Europe if the EU interfered with Russia's 2014 annexation of Crimea (Sasse, 2017; *Daily Mail Online*, 2021). The Obama-Biden Administration in the United States responded by imposing sanctions on Russia.

Russia used its control of the natural gas supply to Europe through Ukraine as geopolitical leverage against EU member states. It showed that the EU did not have a secure energy supply. The desire to attain energy security helped motivate adoption of EU 2050.

Another source of concern for the EU was the Nord Stream 2 pipeline between Russia and Germany. The EU Commission believed that the pipeline once again gave Russia geopolitical leverage over EU foreign policy. This concern was realized when Russia invaded Ukraine in 2022.

The Biden Administration lifted sanctions on the Nord Stream 2 pipeline in 2021. The invasion of Ukraine by Russia on February 24, 2022, motivated the Biden Administration to reimpose sanctions on Russia. Foreign policy analyst Giovanna de Maio observed that "the pipeline's political implications are paramount. Proceeding with Nord Stream 2 risks weakening the credibility of both the EU's sanctions regime against, and overall policy toward, Russia" (de Maio, 2019). The Nord Stream 2 pipeline was apparently sabotaged in 2022, but it is not clear who is responsible (Plucinska, 2022).

Russian annexation of Crimea in 2014 and the Russian invasion of Ukraine in 2022 have demonstrated that energy security should be a crucial policy concern for sovereign states. EU member states have been forced to reassess their dependence on Russian fossil fuels and their timetable for an energy transition.

> **Point to Ponder: How did the 2022 Russian Invasion of Ukraine Affect the Energy Transition in Germany?**
>
> Germany was in the process of implementing the EU 2050 transition to renewable energy when Russia invaded Ukraine on February 24, 2022. As a result of the invasion, the development of the Nord Stream 2 pipeline between Russia and Germany was halted. In addition to discontinuing the pipeline, Germany also decided to restrict installation of more nuclear fission reactors following the March 11, 2011, tsunami and flooding of nuclear fission reactors at Fukushima, Japan.
>
> These energy transition setbacks motivated German officials to embrace German coal resources by reopening coal-fired power plants (Connolly, 2022). In addition, the EU Parliament supported identifying investments in gas and nuclear power plants as climate-friendly (Abnett, 2022).
>
> The EU 2050 timeline for transition to sustainable energy has been affected by the reality of modern war.

13.3 Weaponizing Oil

Crude oil price per barrel has been affected by geopolitical events. In this section, we describe two events that were set in motion by conflicts in the Middle East. We begin by reviewing historical events that are considered the source of turmoil in the Middle East. These events led to weaponization of oil. We first discuss the 1973 Arab oil embargo, and complete the section by describing the Hamas attack on Israel in 2023.

13.3.1 *Events Preceding the First Oil Crisis*

Britain took control of Palestine in 1917 near the end of World War I. When World War I ended in 1918, the British were authorized to administer Palestine by the Mandate of Palestine, which was approved by the League of Nations.

A quarter of a century later, Nazi Germany inflicted a holocaust on Jews in occupied territories during World War II. In 1947, after the end of World War II, Palestine was partitioned by the United Nations into an independent Jewish state, and an independent Arab state. The city of Jerusalem was administered by the United Nations. Jewish Agency Chairman David Ben-Gurion announced the creation of the State of Israel on May 14, 1948, after the departure of the British. Israel became the first Jewish state in 2,000 years with Ben-Gurion serving as Israel's first premier.

Arab leaders rejected the partitioning of Palestine and the creation of Israel. A war between Palestinian Jews and Palestinian Arabs ensued when Israel declared sovereignty. The war escalated to a regional war between Israel and nearby Arab countries. It became known as the Arab-Israeli War. An end to the war was negotiated in July 1949 and the issue that led to the war, the existence of a sovereign Jewish state, was not resolved. The stage was set for a future war in the Middle East that would result in the First Oil Crisis.

The Organization of Petroleum Exporting Countries (OPEC) was founded at a conference in Baghdad, Iraq on September 14, 1960. Thirteen countries became the founding members of OPEC. They included Iran, Indonesia, Angola, Nigeria, Ecuador, Venezuela, and seven Arab countries. The goal of OPEC was to create a bargaining unit for oil-producing countries that had little bargaining power as separate states. The newly formed OPEC gave the members more leverage when being pressured by major oil companies to lower prices.

OPEC sought to exercise more control over the supply of oil and to receive a larger share of oil revenues for its members. It took a few years, but OPEC influence over world oil supply and prices was significant by the early 1970s.

Egypt expelled United Nations forces from the Sinai region in May 1967. The result was a separation of Egyptian and Israeli territory. Israel activated reservists in preparation for armed conflict with Egypt because Israelis felt threatened by the presence of Egyptian forces on their border. Israel attacked Egyptian forces on June 5, 1967. Egyptian forces were joined by Jordanian and Syrian forces in the war against Israel. The war began June 5 and ended on June 10, 1967, with a decisive Israeli victory. It became known as the Six-Day War and resulted in tripling the size of

Israel. As a result of the war, Israel took control of the Sinai Peninsula, the Gaza Strip, the West Bank of the Jordan River including East Jerusalem, and the Golan Heights.

The United States was joined by several other countries a few years after the Six-Day War, in 1971, to stop using gold as a standard for the value of their currencies. As a result of abandoning the gold standard, the value of the dollar and other currencies depreciated. The purchasing power of oil-producing countries was adversely affected by the new currency valuation system.

Price increases in goods sold to oil-exporting countries by oil-importing countries accompanied the change in the value of currencies. Some of the affected goods include wheat, grain, sugar, cement, even refined petroleum products. Members of OPEC and other oil-producing countries received less value from their exported oil even though they were paying higher prices for their goods. OPEC members decided to set the price of oil using the value of gold rather than the value of the dollar. This resulted in a sudden oil price increase that was experienced as an "oil shock" in the United States and other oil-importing countries.

Meanwhile tensions remained high in the Middle East as Arab nations viewed the presence of Israel as an affront to the region. Egypt and Syria attacked Israel on October 6, 1973, the Jewish holiday of Yom Kippur. The two Arab countries wanted to defend their national honor and recover the land they lost in the 1967 Six-Day War. The 1973 war was also known as the Yom Kippur War.

The nations attacking Israel had some success during the first 24 to 48 hours of the war. Although losses were surprisingly high, momentum began to shift in favor of the defending Israelis. The United States, Israel's ally, expected the Yom Kippur War to be short-lived because previous wars did not last long, but the Union of Soviet Socialist Republics (USSR) and Arab allies were sending assistance to Egypt and Syria.

United States Secretary of State Henry Kissinger and President Richard Nixon were concerned about Israeli losses. They launched an airlift operation called Operation Nickel Grass. The operation was designed to supply Israel during the war. The US also provided information to Israel about the location and movement of Egyptian military assets. Israel began a counterattack on October 14, 1973.

Arab nations participating at the Khartoum Arab Summit on September 1, 1967, years before the Yom Kippur War, expressed solidarity against Israel. The Summit participants signed the Khartoum Resolution which included the "three no's": no peace with Israel, no recognition of Israel, and no negotiation with Israel. By supporting Israel in the Yom Kippur War, the United States angered Egypt, Syria, and their Arab allies. They decided to act when President Nixon requested $2.2 billion in emergency aid for Israel from Congress on October 17, 1973.

On October 19, 13 days after the start of the Yom Kippur War, Libya weaponized oil by initiating an oil embargo against the United States. Members of OPEC, under the leadership of King Faisal of Saudi Arabia, followed suit by proclaiming an oil embargo against nations that had supported Israel during the Yom Kippur War.

13.3.2 The 1973 Arab Oil Embargo

The 1973 Arab oil embargo had two goals. The first goal was to incentivize the United States to stop supporting Israel by increasing the price of oil paid by the US and the price of goods that depended on oil within the US. The price of oil was expected to increase when oil imports to the US were discontinued.

The second goal of the embargo was to punish the US for supporting Israel. The US believed that it needed to support Israel because the US and Britain were primarily responsible for the existence of the Israeli state. In addition, since the Soviet Union was supporting the Arab side, the Yom Kippur War could be viewed as a proxy war between the US and the Soviet Union.

The ceasefire for the Yom Kippur War was negotiated primarily by the US and the Soviet Union. The Yom Kippur War ended on October 22, 1973. Hostilities continued for several days and skirmishes continued even longer. The oil embargo did not end when the ceasefire took effect; it continued until March 17, 1974.

The collapse of the US stock market in January 1973 was aggravated by the oil embargo. The 1973–1974 stock market crash lasted from January 1973 to December 1974. The principal cause of the crash was the destabilization of the US currency when President Nixon decided to discontinue the gold standard. The oil embargo contributed to the difficult economic climate because the price of oil increased significantly from January 1973 to January 1974.

Unusually high oil prices forced Americans to ration gasoline. Customers formed long queues at gas stations and many oil companies were accused of price gouging. In the hands of unified oil exporting countries, oil could be used as a weapon to harm the economy of a country, including a global superpower. Oil importing nations around the world realized that oil could be weaponized against them. Many nations began to seriously consider a transition away from a carbon-based economy.

The Arab Oil Embargo is now considered the First Oil Crisis. It is an important example of the interconnection between politics, economics, and energy security. Many other Middle Eastern conflicts have occurred in the decades since the Yom Kippur War. A contemporary Middle Eastern conflict began in 2023 that threatens oil price stability. It is described next.

13.3.3 *The 2023 Hamas Attack on Israel*

Southern Israel was attacked by Hamas on October 7, 2023. Israel's retaliation is a demonstration of how energy can be used as a weapon in war. The context of the conflict is reviewed here by outlining the dispute over control of the region since the creation of the Israeli state in 1948 (Marks, 2023; BOE Gaza Strip, 2023).

Most people living in Palestine before the creation of the Israeli state were Arabs. A sovereign Israeli state was proclaimed on May 14, 1948. Israel was soon at war with five Arab countries: Jordan, Iraq, Syria, Egypt, and Lebanon. The war is known as the 1948 Arab-Israeli War.

Egypt took control of the Gaza Strip after the 1948 Arab-Israeli war. The Gaza strip is a 140 square-mile area located along the Mediterranean Sea between southern Israel and Egypt. Both Palestinians and Israelis have claimed control over the region, and both claim the city of Jerusalem as their capital. Jerusalem has religious and cultural significance to both Arabs and Jews. Figure 13.1 is a map of the region.

Scholars estimate that over 700,000 Palestinians were displaced by the formation of Israel and as a consequence of the 1948 Arab-Israeli War. Many refugees settled in the Gaza Strip while it was under Egyptian control. The Gaza Strip was captured by Israel during the 1967 Six-Day War.

The withdrawal of Israeli forces from the Gaza Strip and other strategic areas was negotiated between Israeli and Palestinian leaders in the Oslo Peace Accords of 1993 and 1995. Israel withdrew from the Gaza Strip in 2005.

Attacks from the Gaza Strip into southern Israel were orchestrated by Hamas on October 7, 2023. Hamas, also known as the Islamic Resistance Movement, is a movement of militant Palestinian nationalists and Islamists in the Gaza Strip and the West Bank. It won an electoral majority

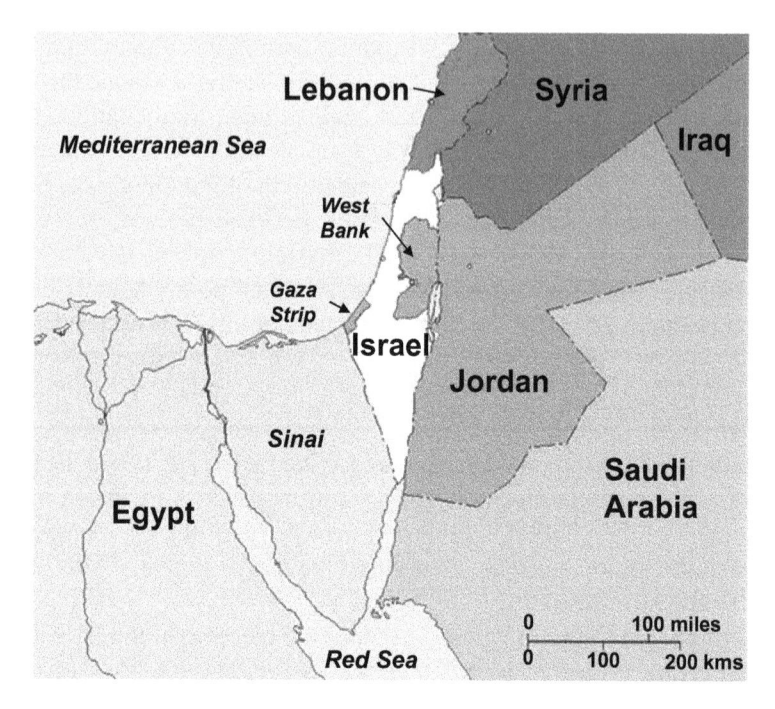

Figure 13.1 Israel and the Gaza Strip

in the Gaza legislature after the 2006 legislative elections in the Gaza Strip. Hamas retained political control over the Gaza Strip when the Gaza legislature dissolved in 2007. The goal of Hamas is to establish an independent Islamic state in historical Palestine.

Israel controls access to the Gaza coastline from the sea. It also controls the northern and eastern borders of the Gaza Strip. The southern border of the Gaza Strip is shared with Egypt. Many experts believe that Hamas weapons, including rockets, are smuggled into the region, or provided by anti-Israeli allies in countries such as Iran.

In 2018, the US Embassy relocated from Tel Aviv to Jerusalem during the Trump Administration. Palestinians interpreted this move as American recognition of Jerusalem as the capital city of Israel. They responded with a protest known as the "Great March of Return" at the Gaza-Israel border. Dozens of protesters lost their lives when Israeli force was used against them.

Tensions continued to rise between Israelis and Palestinians. The October 7, 2023, attack by Hamas resulted in the kidnapping and killing of more than 1,000 Israelis. Victims of the attack included many civilians ranging in age from babies to the elderly. Prime Minister Netanyahu of Israel declared "we are at war" and called for the destruction of Hamas. The Israeli declaration of war against Hamas was motivated by the brutality of the Hamas attack, and it was justified by a history of violent confrontations between Hamas and Israel.

Israel weaponized energy by discontinuing the supply of power and water to the Gaza Strip. The Gaza Strip under Hamas rule did not have its own supply of power and water. Instead, Hamas chose to expand its supply of locally made rockets and built an extensive tunnel system to protect itself from an anticipated retaliation.

Non-combatants in the Gaza Strip were encouraged by Israel to move south toward the border with Egypt but were told to stay in the combat zone by Hamas. Israel began retaliatory airstrikes against Hamas in the Gaza Strip, especially Gaza city, as it prepared to launch a ground offensive. The conflict is ongoing as of this writing.

13.4 Environmental, Social, and Governance — (ESG)

Environmental, social, and governance (ESG) criteria can be used to assess the impact of an organization on its environment, community, and organizational character. Table 13.4 presents several ESG criteria.

Table 13.4 Types of ESG Criteria (Nolan, 2021)

Environmental Issues	Social Issues	Governance Issues
Climate change and carbon emissions	Customer satisfaction	Board composition
Air and water pollution	Data protection and privacy	Audit committee structure
Biodiversity	Gender and diversity	Bribery and corruption
Deforestation	Employee engagement	Executive compensation
Energy efficiency	Community relations	Lobbying
Waste management	Human rights	Political contributions
Water scarcity	Labor standards	Whistleblower schemes

Measures to avoid conflicts of interest and unethical behavior may be included in ESG criteria. Social criteria can include diversity, equity, and inclusion (DEI) metrics as a means of supporting a sustainable workforce. For example, an organization can respond to government compliance requirements by establishing an office of Diversity and Inclusion to promote inclusion of diverse stakeholders, and an office of Equity to address reports of alleged discrimination and harassment. In this case, an alternative acronym would be DI&E, rather than DEI, which is the Latin word *Dei* for deities and could be considered offensive.

ESG criteria can be used to construct metrics that assess how well a company performs relative to the metrics. ESG metrics have not been standardized. They are intended to motivate businesses to think about how their actions affect society and the environment. It should also be possible to maximize financial performance using ESG metrics. An assessment of an organization with respect to ESG metrics can be conducted by the organization itself or by a third party.

Socially responsible investing (SRI) is exemplified by the integration of ESG assessments into investment decision-making. SRI originated in the 1960s and 1970s when people attempted to use investing to advance political causes like ending the Vietnam War, advancing civil rights, raising public awareness of environmental issues, and combating apartheid in South Africa.

Today, ESG metrics can supplement the use of more traditional economic factors to analyze investment decisions. Investors should also recognize that ESG metrics can represent a political perspective. Examples of the political application of ESG metrics include designing metrics that encourage investment in renewable energy systems or discourage financing of politically undesirable projects such as fossil fuel development.

13.5 China

China is playing an increasingly important global role under the leadership of Xi Jinping. Xi began serving as general secretary of the Chinese Communist Party in 2012 and assumed the position of President of the People's Republic of China in 2013. Here we discuss two Chinese initiatives: Xi's Belt and Road Initiative (BRI), and a new multinational organization known as BRICS.

13.5.1 *China's Belt and Road Initiative*

Xi Jinping began China's BRI in 2013. He "proposed building a land-based 'Silk Road Economic Belt,' extending from China to Central and South Asia, the Middle East, and Europe, and a sea-based '21st Century Maritime Silk Road,' connecting China to Southeast Asia, the Middle East, Africa, and Europe via major sea lanes (Figure 13.2). Together, these would form the Belt and Road Initiative (BRI), still known officially in Chinese as 'One Belt, One Road'" (CFR BRI, 2021, p. 8). BRI is considered Xi's signature foreign policy initiative.

The geographic extent of the BRI is sketched in Figure 13.2. The Belt approximately follows the ancient Silk Road between China and Europe. Land-based routes from Xi'an, China to Rotterdam, Holland pass through Central Asia. The Road from Fuzhou, China to Venice, Italy follows sea-based routes through Southeast Asia, Africa, and the Middle East.

The Council on Foreign Relations (CFR BRI, 2021, p. 2) says that "BRI was initially designed to connect China's modern coastal cities to its underdeveloped interior and to its Southeast, Central, and South Asian

'Belt' connects land routes
'Road' connects ocean routes

Figure 13.2 Sketch of China's BRI

neighbors, cementing China's position at the center of a more connected world." Projects ranging from "power plants, railways, highways, and ports to telecommunications infrastructure, fiber-optic cables, and smart cities around the world" (CFR BRI, 2021, p. 2) have been financed and built by Chinese institutions.

The original scope of BRI has expanded to include projects in over 100 countries. An activity is considered a BRI project and is typically identified by formal agreements with China. Projects include "a Digital Silk Road intended to improve recipients' telecommunications networks, artificial intelligence capabilities, cloud computing, e-commerce and mobile payment systems, surveillance technology, and other high-tech areas, along with a Health Silk Road designed to operationalize China's vision of global health governance" (CFR BRI, 2021, p. 2).

China is using BRI to develop strategic economic relationships with other nations (Chandra, BRI, 2022). Financial institutions and businesses are being encouraged by China to enhance cross-border telecommunications; build international transport corridors using roads, bridges, and deepwater ports; and construct international energy infrastructure such as oil and gas pipelines, and electricity transmission lines across borders.

Some analysts are concerned about China's increasing influence in Southeast Asia (Kurlantzick, 2022). Examples of Chinese activities that concern China observers are listed in Table 13.5.

Construction of military installations on islands and reefs in the South China Sea is covered by Item D in Table 13.5. China can use the presence of military installations to support territorial claims in the

Table 13.5 Examples of Concerning Behavior by China

Item	Behavior
A	Police state intervention to pacify Hong Kong
B	Authoritarian lockdowns during Covid in major cities such as Shanghai
C	An apparent willingness to use economic coercion against countries participating in BRI agreements
D	Military expansion in the South China Sea
E	Growing ties with Russia

South China Sea, which is thought to contain such valuable natural resources as oil and gas.

Item E in Table 13.5 refers to the strengthening of Chinese relations with Russia before the 2022 Russian invasion of Ukraine. China and Russia have maintained ties throughout the Russia–Ukraine war. China's willingness to buy Russian oil has weakened sanctions imposed on Russia by the United States and the European Union.

Kurlantzick (2022) believed that leading democracies might be able to resolve global issues such as public health and anthropogenic climate change by cooperating with China. He realized that China was attempting to control foreign media, academia, and other hubs of influence. For example, China could influence the policies of foreign institutions by providing covert funding. China could also infiltrate and control Chinese communities in other countries. Members of the emigrant community could be influenced by threatening extended family members that were still in China.

CFR President Richard Haas presented a different view of China's role in BRI (CFR BRI, 2021, Foreword). Haas wrote that some analysts believed that "BRI is China's primary conduit for pursuing global domination. This Task Force, however, argues that China pursued BRI primarily to address a number of domestic issues... Much of Chinese foreign policy is animated by a drive to bolster domestic political stability."

13.5.2 *Brazil, Russia, India, China, and South Africa — BRICS*

BRICS is a multinational organization named after its five member states: Brazil, Russia, India, China, and South Africa (CFR BRICS, 2017; Investopedia, 2023). It began when leaders of China, India, and Russia met at a 2008 meeting of the Group of 8 (G8) in St. Petersburg, Russia. The G8 was formed in 1997 when Russia was added to the Group of 7 (G7) member countries. G7 members included the United States, the United Kingdom, France, Germany, Italy, Canada, and Japan. The G8 existed from 1997 until Russia was expelled in 2014 following the Russian seizure of Crimea from Ukraine.

China, India, and Russia met with Brazil in 2009 and became known as BRIC. South Africa was added to the group at its 2010 meeting to form BRICS. BRICS could be viewed as a representation of emerging economies around the world. It was initially interested in achieving a greater role in global governance for countries with emerging economies.

China and Russia already played a significant role in global governance because they occupied two of the five permanent seats on the UN Security Council. The other three permanent seats were held by the United States, the United Kingdom, and France.

The member countries of BRICS are not a homogenous group. China and Russia tend to be authoritarian, while Brazil, India, and South Africa are more correctly characterized as democracies. The populations of China and India exceed a billion people each, while Brazil, Russia, and South Africa have much smaller populations. China and India also share a common border that has led to tension between the two countries.

BRICS has "expanded its diplomatic activities, advocated a larger voice in global economic and security forums for its members, and created brand new financial institutions" (CFR BRICS, 2017). In addition to geopolitical influence, the goals of BRICS member nations include economic cooperation and expanding multilateral trade and development. Many countries have expressed interest in joining BRICS because the group is considered an "alternative to global bodies viewed as dominated by the traditional Western powers and hope membership will unlock benefits including development finance, and increased trade and investment" (Reuters BRICS, 2023). In January 2024, the following five countries joined the original five BRICS countries: Saudi Arabia, Egypt, United Arab Emirates, Iran, and Ethiopia.

Point to Ponder: Is the World Developing Two Financial Systems with Two Global Currencies?

It appears that BRICS is pursuing the development of a new world order with a new global currency. A global currency is created and sustained by a central bank and is intended for use in all transactions worldwide. The United States dollar (US$) has functioned as the global currency since World War II because of the relative strength of the US economy when compared with other national economies. When the US imposes sanctions on other countries, it often uses access to the US$ and the international financial system as a tool for influencing the behavior of other countries. BRICS is challenging the role of the US$ as the global currency.

Chapter 14

Weaponizing a Pandemic

In this chapter, we review the historical steps that led to the unionization of public sector employees. We then present a timeline of key events that occurred during the COVID-19 Pandemic. The pandemic helped expose the politicization of public sector employees. A specific example is presented to illustrate the politicization: the 2020 Presidential Election in the United States. Civil rights abuses that can occur when a pandemic is weaponized are then described.

14.1 Unionizing Public Sector Employees

The relationship between presidential policies and unions can influence union politics. Political action committees operated by government unions can accept voluntary contributions and make donations to political campaigns. As a result, public sector unions can impact elections on the local, state, and national levels. Here, we summarize the unionization of Federal government employees.

14.1.1 *History of Public Sector Unions*

Rutgers University labor economist Leo Troy describes public sector unionism as the New Socialism (Troy, 1994). Old Socialism supported state control of the means of production and distribution. A new system appeared after the New Deal. The administration of Franklin Delano Roosevelt enacted the New Deal, a domestic initiative, from

1933 to 1939. The New Deal sought to reform such sectors as housing, labor, hydropower, industrial, agricultural, and banking.

The scope of the Federal Government was greatly expanded by the New Deal. The new system in the United States has been called private socialism. In this system, public sector unions have the power to elect their employers and bargain with them. A good example is the election of school board members by public school unions. Public sector unions can vote for politicians who can enact legislation that would give the unions more power to increase their income. An example of such legislation is the Wagner Act.

The Wagner Act can be used to demarcate three historical periods: "the employment-at-will period before the Wagner Act of 1935; the Wagner Act period of the mid-twentieth century; and the public union period since 1958" (Morena, 2011, p. 1). During the employment-at-will era, the common law labor relations could be characterized as "master and servant," evoking the feudal system in Europe during the Middle Ages. The majority of historians believe that the 19th century marked the end of the premodern system and the introduction of employer–employee relations that are suitable for a contemporary, democratic, egalitarian society. Slavery was abolished in the U.S. after the Civil War (1861–1865). This enabled the creation of a legal system that treated everyone, in principle, as equal under the law. Relations between employer and employee were expected to be voluntary and contractual.

Groups of workers were allowed to organize labor unions. Workers could strike, that is, quit as a group, to achieve their goals. By contrast, employers were allowed to replace workers who quit. To stop strike-busting employers from conducting business, unions have at times employed violence and threats against replacement workers or sabotaged their places of employment. The law was authorized to intervene to maintain order and protect non-striking workers and employers.

The "free labor" system was supposed to be a system of freedom to enter contracts. Labor leaders claimed that employers had all the power to set wages in the "free labor" system because it was a coercive system. The idea that unions were needed to organize workers was bolstered by the expansion of private sector enterprises. The old employment-at-will doctrine was replaced by a system of labor law that gave unions special privileges. The National Labor Relations Act of 1935, also known as the Wagner Act, was based on the belief that legislation was needed to balance the bargaining power of labor and management.

According to the Wagner Act, employers were obligated to engage in collective bargaining with any group selected by the majority of its employees. The Wagner Act was modified by the Taft–Hartley Act in 1947. Under the Taft–Hartley Act, states were allowed to enact right-to-work laws. Industries began to move from Wagner Act, or union-shop, states to right-to-work states.

Public employees were specifically excluded from coverage by the Wagner Act. In a 1937 letter to Luther Steward, the president of the Federation of Federal Employees, President Franklin D. Roosevelt explained that "All government employees should realize that the process of collective bargaining, as usually understood, cannot be transplanted into the public service… The very nature and purposes of government make it impossible for administrative officials to represent fully or to bind the employer in mutual discussions with government employee organizations. The employer is the whole people, who speak by means of laws enacted by their representatives in Congress. Accordingly, administrative officials and employees alike are governed and guided, and in many instances restricted, by laws which establish policies, procedures, or rule in personnel matters" (Moreno, 2011, p. 4). The government cannot be compelled to bargain collectively because it is sovereign.

Membership in public unions was voluntary. The primary goal of public unions was to improve working conditions. Presidents Theodore Roosevelt (1901–1909) and William Howard Taft (1909–1913) recognized that Federal employee organizations could exercise undue influence on Congress by lobbying Congress. Roosevelt and Taft issued executive orders that prohibited Federal employees from joining Federal employee unions.

Congress overturned the executive orders issued by Roosevelt and Taft when it passed the 1912 Lloyd–La Follette Act. One of the sponsors of the act was Senator Robert La Follette from Wisconsin, the first state to support public employee unions. The Lloyd–La Follette Act gave public employee organizations the right to petition Congress. It did not authorize collective bargaining or the right to strike.

The 1919 Boston police strike dealt a major blow to public sector unionism because the public did not support the strike. Calvin Coolidge became a national hero when he declared that "There is no right to strike against the public safety by anybody, anywhere, anytime" (Moreno, 2011, p. 5).

Public opinion was more receptive to public sector labor unions in the late 1950s. Public employees in Wisconsin were granted the right to organize and bargain collectively, but not strike, after Democrats took control of Wisconsin in the 1958 elections.

President John F. Kennedy authorized federal employees to form unions and bargain collectively when he signed Executive Order 10988 (EO10988) in 1962. EO10988 did not allow unions to bargain over wages, nor did it allow unions to strike or compel Federal employees to join a union. President Richard Nixon strengthened EO10988, and it was given statutory status when the Civil Service Act of 1978 was enacted. The authorization of public sector unions on the Federal level is discussed in more detail in the following section.

14.1.2 *Authorization of Public Sector Unions*

On January 17, 1962, President John F. Kennedy issued Executive Order 10988: Employee Management Cooperation in the Federal Sector. On June 22, 1961, Kennedy sent a memorandum creating EO10988 to the heads of all executive departments and agencies. Kennedy said in the memorandum that "The participation of employees in the formation and implementation of employee policy and procedures affecting them contributes to the effective conduct of public business" (FLRA EO10988, 2023).

A Task Force on Employee Management Relations in the Federal Service endorsed President Kennedy's view that "the public interest calls for strengthening of employee management relations within the Federal Government... Certain categories of Federal employees very much want to participate in the formulation of personnel policies and have established large and stable organizations for this purpose." Federal employees were granted the right to form, join, or assist labor organizations by EO10988.

A Presidential Review Committee on Employee Management Relations in the Federal Service was established by President Lyndon B. Johnson in 1967 to review the previous five years of experience with EO10988. President Nixon expanded the rights granted by EO10988 in 1969 by implementing the Presidential Review Committee's recommendations. Nixon issued EO11491 to establish "an institutional framework to govern labor-management relations in the Federal Government, set forth specific unfair labor practices, and authorized the use of binding arbitration of certain disputes" (FLRA EO10988, 2023).

Table 14.1 Responsibilities of the Council and Panel (FLRA EO10988, 2023)

Entity	Responsibilities
Council	Oversee the program
	Make definitive interpretations and rulings on provisions of EO11491
	Decide major policy issues
	Hear appeals, at its discretion, from decisions made by the Assistant Secretary of Labor for Labor-Management Relations on unfair labor practice charges and representation claims
	Resolve appeals from negotiable decisions made by agency heads
	Decide exceptions to arbitration awards
Panel	Has discretionary authority to assist parties in resolving bargaining impasses when voluntary arrangements failed

EO11491 established two new entities: the Federal Labor Relations Council (Council) and the Federal Service Impasses Panel (Panel). The responsibilities of the Council and Panel are summarized in Table 14.1.

In 1975, President Gerald R. Ford amended Nixon's EO11491 by issuing EO11838. Ford's EO11838 expanded collective bargaining rights to include agency regulations and mid-contract changes, enhanced resolution procedures for third-party disputes, and directed unions to recognize secret ballot elections.

President Jimmy Carter decided in 1977, the year he took office, that the civil service system needed comprehensive reform. The Civil Service Reform Act was passed in 1978. It included Title VII, which created the Federal Labor Relations Authority (FLRA). Carter signed Title VII: the Federal Service Labor Management Relations Statute (Statute) as part of the Civil Service Reform Act on October 13, 1978. The Statute became effective January 11, 1979. It recognized that organized labor and collective bargaining in the civil service were in the public interest. Some of the items in the Statute included prohibiting employees from striking, establishing an independent Office of the General Counsel to investigate and prosecute unfair labor practice charges, and making the FLRA final orders subject to judicial review.

In 1993, President William J. Clinton issued EO12871: Labor Management Partnerships that called for Federal labor and management to serve as partners in designing and implementing comprehensive changes to Federal government agencies that could improve their ability to deliver high-quality services to the American people. President George W. Bush rescinded Clinton's EO12871 with EO13203: Revocation of Executive Order and Presidential Memorandum Concerning Labor-Management Partnerships.

In 2009, President Barack Obama issued EO13522: Creating Labor-Management Forums to Improve the Delivery of Government Services. The purpose of EO13522 was to implement a form of labor-management relations throughout the Executive Branch that would encourage management to collaborate with labor to develop joint solutions to workplace issues.

Jessie Bur reported that "Each presidential administration takes a different approach to union relations. President Donald Trump issued executive orders that restricted union activities and prescribed collective bargaining objectives for agency operations. On the other hand, the Biden administration has taken an openly pro-union approach, rescinding the previous orders and instructing agencies to include unions in many forms of decision-making, even those not required by law" (Bur, 2021).

Former federal government official Kash Patel wrote that "Congress should ban the existence of public sector unions, which, by their nature, establish an antagonistic relationship between federal bureaucrats and the administration they are employed to work for … Those who do work for the executive branch should know that they serve at the executive branch's discretion. They should not be able to leverage the power of a union to win concessions from their employer, the president of the United States" (Patel, 2023, p. 176).

Most Federal union contributions are directed to presidential candidates who support the growth of government and unions. In the United States, unions tend to favor Democrat candidates rather than Republican candidates.

14.2 The 2020 COVID-19 Pandemic

The history of Coronavirus disease 2019 (COVID-19) is reviewed in this section. We then describe relevant cell biology, highlight key moments during the pandemic, and discuss the abuse of civil rights that occurred during and after the pandemic.

14.2.1 *History of COVID-19*

The Mayo Clinic reported that Harvard University researchers used a synthesized ribonucleic acid (RNA) enzyme in 1984 to make biologically active messenger RNA (mRNA) in a laboratory (Mayo Clinic COVID-19, 2023). Synthetic mRNA can be used to study the activity and function of genes.

A few years later, in 1987, American physician and biochemist Robert W. Malone formed a mixture of mRNA with droplets of fat. He then added human cells to the mixture and discovered that the human cells absorbed mRNA and made proteins. He also found that frog embryos could absorb the mRNA. Malone's experiments are considered early steps in the eventual development of mRNA-based COVID-19 vaccines.

Researchers experimented with mRNA in rats and mice during the 1990s. For example, they attempted to use mRNA to develop influenza and cancer vaccines in mice. The search for treatments and vaccines using mRNA continued in the 2000s. The research was hampered by the cost of producing mRNA and its susceptibility to damage.

In 2005, Katalin Kariko and Drew Weissman discovered that synthetic mRNA can be modified so that it is not attacked by the immune system. Researchers began studying mRNA in the 2010s. They were interested in determining how mRNA could be used as a vaccine or treatment for disease.

A new disease called COVID-19 appeared in 2019 as a possible application of mRNA vaccination technology. The COVID-19 disease can be spread by droplets and is caused by severe acute respiratory syndrome Coronavirus 2 (SARS-CoV-2). The World Health Organization (WHO) declared the COVID-19 outbreak a pandemic in 2020.

The development of an mRNA vaccine to combat COVID-19 requires an understanding of key elements of cell biology, which is presented in the following primer.

14.2.2 *Primer on Cell Biology*

Every living system consists of cells. The cell wall, or cell membrane, bounds the chemical contents of each cell. The chemical contents include large, complex chemicals known as proteins and nucleic acids.

Nucleic acids and proteins are two important types of biochemicals. Proteins are assembled from smaller molecules called amino acids. Each protein is a long chain of amino acids. One type of protein is called an enzyme. It can be used in the cell to catalyze a chemical process without being changed by the process. Enzymes are needed to replicate nucleic acids.

All known life forms depend on two types of nucleic acids: deoxyribonucleic acid (DNA) and ribonucleic acid (RNA). DNA consists of a chain of four molecules bound to deoxyribose phosphate, a sugar base. The shape of the large DNA molecule is a double helix. Each helical strand is comprised of deoxyribose phosphate molecules bound together to form a chain. Organic molecules called purines and pyrimidines are attached to the helical strand. The purines are adenine (A) and guanine (G), and the pyrimidines are thymine (T) and cytosine (C). A purine and a pyrimidine can combine to form a base pair that connects the two helical strands of a DNA double helix. Figure 14.1 illustrates DNA and its components.

The building blocks of nucleic acids (DNA and RNA) are called nucleotides. In humans, the organic molecules adenine (A), guanine (G),

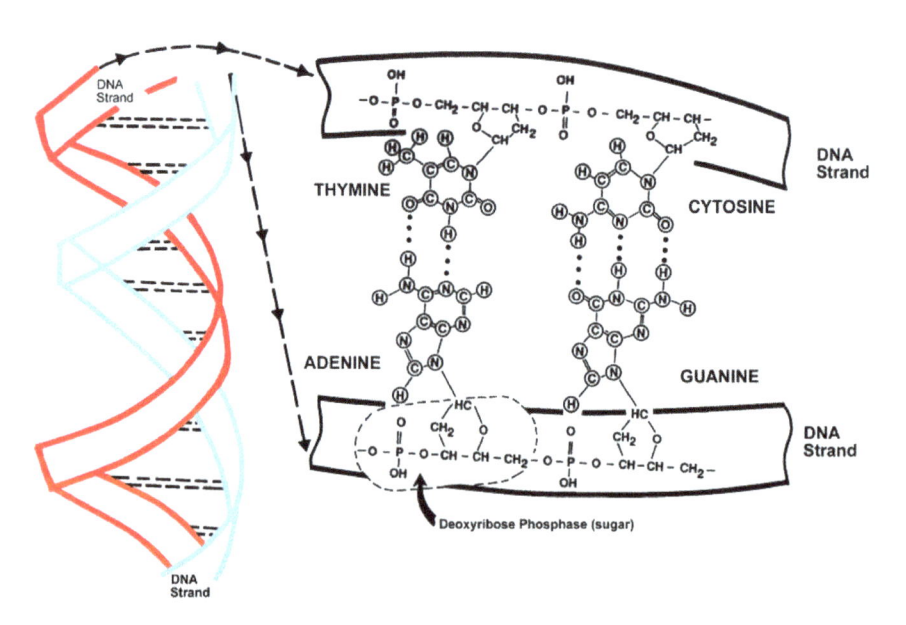

Figure 14.1 Diagram of DNA and Its Components

thymine (T), and cytosine (C) are nucleotides. A base pair in human DNA is either an adenine–thymine (A–T) pair or a guanine–cytosine (G–C) pair. The rungs of the DNA double helix are formed by DNA base pairs. The sequence of DNA base pairs that contains complete genetic information for an organism is a genome.

Nucleic acid in the cell tells the cell what to do and controls reproduction. Protein synthesis in the cell occurs when the genetic code carried by nucleic acids is translated into proteins by ribosomes in the cell. Protein-synthesizing ribosomes contain RNA.

The RNA molecule has a single helical structure. It is the template for DNA replication when a cell divides. RNA is also involved in the synthesis of cellular proteins. The instructions in the DNA molecule are used to synthesize cellular proteins. The information contained in DNA cannot be used directly in protein synthesis; it must be communicated through the intermediate RNA.

A chromosome consists of a protein and a single DNA molecule. Each chromosome is a slender structure inside the nucleus of animal and plant cells. The human genome consists of 23 pairs of chromosomes. Genes are found at specific sites on a chromosome. The gene is the basic unit of heredity. It is a segment of DNA that contains instructions for synthesizing a particular protein.

The central dogma of molecular biology is expressed in terms of the relationships between DNA, RNA, and proteins. The transmission of genetic information flows from DNA to RNA to protein, or from RNA directly to protein. The information flow path from DNA to RNA to protein says that DNA is first copied by RNA in the replication process; genes are then read and converted into mRNA in the transcription process; lastly, the mRNA is turned into a protein by ribosomes in the translation process. RNA is used as an intermediary in DNA replication.

A practical application of the above information in the context of the COVID-19 Pandemic is that viruses have only DNA or RNA; therefore, they can only reproduce with the help of a host organism. By contrast, the presence of both DNA and RNA in bacteria means that bacteria can reproduce without other organisms.

Mutations, or errors in DNA replication, can occur. These errors are required for changes to occur in a unicellular species. New combinations of genetic material are routinely created by more complex, sexually active organisms. These new combinations, as well as errors in the accurate replication of combinations, are mechanisms that enable the evolution of sexual

Table 14.2 Global Deaths Associated with COVID-19 Infections

As of date	Number
April 10, 2020	100,000
May 15, 2020	300,000
June 28, 2020	500,000
September 28, 2020	1 million
January 15, 2021	2 million
Early April, 2021	3 million
July 7, 2021	4 million
November 1, 2021	5 million

organisms. Asexual organisms do not have the same mechanisms. Mutations are required for asexual unicellular organisms to change and evolve.

14.2.3 *The COVID-19 Pandemic*

The increase in the number of worldwide deaths associated with COVID-19 infections is summarized in Table 14.2. It shows how rapidly the death toll was rising and explains the concern felt by many public health officials.

Key moments during the COVID-19 Pandemic are presented as follows in the context of a timeline that focuses on 2020 and extends into 2021 (Katella, 2021; Whiting and Wood, 2021).

December 2019
- In early December 2019, the first pneumonia-like symptoms of a new illness, soon to be known as COVID-19, appeared. According to some sources, similar symptoms appeared in November 2019. The disease could be passed from person to person.
- On December 21st, the WHO was notified of a cluster of pneumonia cases in the city of Wuhan, Hubei Province, China.

January 2020
- In early January 2020, the Huanan seafood market was closed because it appeared to be connected to the new disease.

- On January 20th, the COVID-19 virus appeared in the United States. The first confirmed case was a man who had returned from Wuhan.
- To stop the spread of the disease, a lockdown was imposed on Wuhan by Chinese authorities until January 23rd. A lockdown is a policy that requires a group of people to remain in one location, usually due to specific risks that could harm the people if they moved and interacted freely.
- On January 30th, the WHO Director-General declared the outbreak a Public Health Emergency of International Concern. The Trump administration barred any foreign national who had visited China in the last 14 days from entering the United States.
- The campaign for President of the United States was entering the primary season.

February 2020
- On February 2nd, the first coronavirus death outside of China was reported: a man in his 40s from the Philippines.
- On February 11th, COVID-19 was formally named by WHO. The virus responsible for the illness is known as SARS-CoV-2, according to the International Committee on Taxonomy of Viruses.
- On February 23rd, two hotspot areas in Italy were under lockdown. Other nations implemented lockdowns shortly thereafter.
- On February 28th, Nigeria reported the first case in Sub-Saharan Africa.

March 2020
- On March 24th, the lockdown in Hubei Province, China was partially lifted.
- The 1.3 billion people in India were placed under lockdown by the Indian prime minister.
- The Tokyo Olympics in Japan was postponed until 2021.
- California was the first state to order residents to stay home.
- As the number of cases rose, hospitals worried about a national shortage of personal protective equipment (PPE).
- On March 27th, Trump signed a $2 trillion stimulus package aimed at resolving the economic and health crises.

April 2020
- Countries kept their borders closed. Job losses increased as businesses shut down. Some schools closed, sporting events were canceled, and

college students went home. People began wearing masks and maintaining a distance of 6 feet (or 2 m) from other individuals in a practice called "social distancing."
- Approximately 10 million Americans applied for unemployment benefits because of job losses associated with the disease.
- On April 6th, the WHO estimated that school closures affected up to 90% of students.

May 2020
- Experts attempted to flatten the curve. Flattening the curve was a public health strategy to slow down the spread of the SARS-CoV-2 virus during the early stages of the COVID-19 Pandemic. The epidemic curve is the curve being flattened. It is a visual representation of the number of infected people needing health care over time. The goal of flattening the curve was to prevent the healthcare system and infrastructure from being overwhelmed by reducing the rate of infection.
- On May 1st, the U.S. Food and Drug Administration (FDA) issued emergency approval for Remdesivir, a hepatitis and Ebola treatment, to be used to treat COVID-19.
- States started a phased reopening based on criteria provided by the Trump Administration following months of lockdown.

June 2020
- New cases appeared when attempts were made to reopen the economy. The epidemic curve was not flattening.

July 2020
- The pandemic and social reactions to the pandemic, such as lockdowns and closures, caused an increase in mental health problems.
- Moderna, a U.S. firm, was the first to start a human trial of a COVID-19 vaccine. It published early results that an immune response against the virus was triggered by its vaccine. It said that there were no serious side effects at the time.

August 2020
- The first documented case of someone being reinfected was reported in Hong Kong.

September 2020
- The school year opened with a variety of strategies designed to keep students and teachers safe. Some strategies included in-person classes with appropriate protective equipment such as masks, remote classes using online internet software, and hybrid classes that could include a combination of in-person and remote learning.

October 2020
- The U.S. FDA granted full approval for Remdesivir to treat COVID-19.

November 2020
- Cases increased as people spent more time inside to avoid colder weather. As the holidays approached, the Center for Disease Control and Prevention (CDC) encouraged Americans to avoid interacting with people who visited their homes, keep gatherings small, and stay at home.
- Pfizer created a vaccine candidate with BioNTech and announced the results of the first interim efficacy analysis. The results showed that the vaccine candidate was more than 90% effective in preventing COVID-19 in participants who did not have evidence of SARS-CoV-2 infection previously.
- On November 16th, Moderna announced that the vaccine's Phase 3 trial results indicated an efficacy of over 94%.
- Other vaccine candidates were under development around the world.

December 2020
- The FDA gave the first emergency use authorization (EUA) for an mRNA vaccine to Pfizer-BioNTech. Moderna received an EUA for another mRNA vaccine a week later.
- Major variants of the virus were beginning to appear. The WHO announced that the first variant of concern was documented in South Africa. The variant was called the B.1.351 variant and is known as the Beta variant.
- On December 2nd, the United Kingdom became the first nation to approve the Pfizer-BioNTech COVID-19 vaccine. Two days later, Bahrain approved the vaccine.

Table 14.3 Selected COVID-19 Pandemic Highlights in 2021

Date	Comment
February	The G7 countries pledged $4.3 billion to support worldwide access to testing, therapies, and vaccines. The G7 commitment was part of a total commitment of $10.3 billion.
March	Johnson & Johnson's COVID-19 vaccine was listed by the WHO for use in emergencies following approval from the European Medicines Agency.
March 30	According to the report, it was "highly likely" that the virus was transmitted to humans by a bat.
June	The list of vaccines approved by WHO increases to six.
July 1	The EU launched its digital COVID-19 certificate. The certificate confirmed the vaccination status of EU citizens. EU citizens were supposed to be vaccinated to move between EU member states.
September	Johnson & Johnson released results from a real-world study that showed its COVID-19 vaccine had a 74% efficacy rating.
November	A study showed that a COVID-19 antiviral pill developed by Pfizer reduced the risk of hospitalization or death by 89%.

- The WHO issued emergency use validation for the Pfizer-BioNTech COVID-19 vaccine. It was the first COVID-19 vaccine to receive validation from the WHO for emergency use.

Table 14.3 presents a few highlights from 2021.

14.3 Abuse of Civil Rights

The reaction of society to the COVID-19 Pandemic exposed the fragility of Western Civilization. We display the fragility by considering three specific examples: the 2016 Presidential election in the United States; the 2020 Presidential election in the United States during the pandemic; and examples of the international response to the pandemic.

Point to Ponder: What is a Vaccine?

It was standard practice during the COVID-19 Pandemic to call the medicines developed to combat the COVID-19 illness vaccines. Many vaccines are made from inactivated microorganisms, while others consist of biological materials such as antibodies, white blood cells called lymphocytes, or mRNA. They are administered to prevent disease. The history of the use of mRNA medicines has shown that they do not prevent COVID-19 infections or reinfections, but mRNA medicines are thought to reduce the impact of COVID-19 infection. Some undesirable cardiovascular side effects are associated with mRNA medicines (Yasmin et al., 2023).

14.3.1 *The 2016 Presidential Election*

Republican Donald Trump defeated Democrat Hillary Clinton in the 2016 Presidential election. Trump was considered a surprise winner because he was a billionaire businessman without experience in elective office. He defeated several Republican challengers, including Governors and Senators, during the 2016 Primary elections. After becoming the Republican Presidential nominee, he faced Democrat Presidential nominee Hillary Clinton. Clinton had been a First Lady in Arkansas, First Lady in the White House, U.S. Senator from New York, and Secretary of State during the Obama-Biden Administration.

The Federal Bureau of Investigation (FBI) became publicly involved in the 2016 election because of its Director James Comey (BOE Comey, 2023). Comey began working as an assistant U.S. attorney for the Southern District of New York in 1987. He was appointed by Rudy Giuliani, who was appointed U.S. attorney for the Southern District of New York in 1983 (BOE Giuliani, 2023). Giuliani was elected Mayor of New York City in 1993 and was Mayor on September 11, 2001, when terrorists flew two airliners into the Twin Towers buildings.

Robert Mueller retired as FBI Director and was replaced by James Comey in June 2013. Comey was appointed by Democrat President Barack Obama in 2013. Comey started his 10-year term on September 4, 2013, after he was confirmed by the U.S. Senate. His FBI investigated Hillary Clinton's use of a private email server while she served as Secretary of State.

In a surprising move, Comey questioned Hillary Clinton's judgment during a news conference on July 5, 2016, just weeks before she was nominated as the Democratic Presidential nominee on July 28, 2016. He recognized that "Although there is evidence of potential violations of the statutes regarding the handling of classified information, our judgment is that no reasonable prosecutor would bring such a case" (FBI Comey Remarks, 2016). Comey pointed out that a review of FBI investigations into the mishandling or removal of classified information from authorized locations, the FBI could not "find a case that would support bringing criminal charges on these facts. All the cases prosecuted involved some combination of: clearly intentional and willful mishandling of classified information; or vast quantities of materials exposed in such a way as to support an inference of intentional misconduct; or indications of disloyalty to the United States; or efforts to obstruct justice. We do not see those things here" (FBI Comey Remarks, 2016).

To avoid any misunderstanding, Comey said that "this is not to suggest that in similar circumstances, a person who engaged in this activity would face no consequences. To the contrary, those individuals are often subject to security or administrative sanctions. But that is not what we are deciding now.

As a result, although the Department of Justice makes final decisions on matters like this, we are expressing to Justice our view that no charges are appropriate in this case" (FBI Comey Remarks, 2016).

Donald Trump, the Republican Presidential nominee, and other opponents of Hillary Clinton strongly criticized the decision not to recommend charges. The investigation was essentially reopened by Comey less than two weeks before the election on November 8 when he disclosed his review of recently discovered Clinton emails in a letter to Congress. Two days before Clinton lost the election, Comey announced that no illegal activity had been found in the newly discovered emails.

Comey was reportedly asked to remain as FBI director by President Trump after Trump took office in January 2017. Trump claimed in March that Obama had wiretapped his phones during the 2016 campaign, but Comey dismissed Trump's claims in testimony to a congressional committee later in March. Comey acknowledged that the FBI had been investigating allegations of collusion between Trump campaign members and Russian officials during the 2016 election. In another report, CNN said that "U.S. investigators wiretapped former Trump campaign chairman Paul Manafort under secret court orders before and after the 2016 election" (Reuters Wiretapping, 2017). Wiretapping did occur during the Obama Administration as well as during the Trump Administration.

Trump fired Comey on May 9th, citing DOJ officials who said Comey had treated Clinton unfairly. The FBI's handling of the Clinton case was the subject of an investigation overseen by the DOJ Inspector General. Results of the investigation were published in June 2018. The report concluded that Comey's departures from standard FBI procedures harmed the agency's reputation for impartiality and called Comey's actions insubordinate. Furthermore, the report found no evidence that Comey was politically motivated, nor did it criticize the FBI for not choosing to prosecute Clinton.

Comey's dismissal was criticized for allegedly being related to the Russia investigation. Trump later acknowledged that Comey was fired in part because of the Russia investigation.

A memo written by Comey was leaked to the media in May 2017. The memo about a February meeting between Comey and Trump claimed that the president had instructed him to end the FBI investigation into Michael T. Flynn, the former national security adviser to President Trump. Flynn was under investigation as part of the Russia investigation. Former FBI Director Robert Mueller was appointed special counsel to the Russia investigation shortly after the memo was made public.

Comey testified before the Senate Intelligence Committee in June 2017. He said that he had written memos documenting his conversations with Trump because he was concerned that the president might misrepresent the content of a conversation. In addition, Comey disclosed that he had subtly leaked the memo in May 2017 with the objective of getting a special counsel appointed to oversee the Russia investigation. Comey said the FBI was not investigating President Trump.

14.3.2 *The 2019 Donald Trump Impeachment*

The first impeachment of Donald Trump in 2019 illustrates the partisan culture in the U.S. Congress. We begin with the completion of Special Counsel Robert Mueller's investigation into Russian interference in the 2016 election and possible collusion between the Trump campaign and the Kremlin (Beavers *et al.*, 2019). Mueller presented his report to U.S. Attorney General William Barr on March 22, 2019. According to a summary of Mueller's report released by Barr, Mueller's investigation did not find evidence that the Trump campaign conspired or coordinated with the Russian government to interfere in the 2016 election. Barr also wrote that he reviewed Mueller's findings with Deputy Attorney General Rod Rosenstein and decided not to pursue an obstruction of justice charge against Trump.

Democrats demanded the release of the complete Mueller report while Republicans declared that Trump had been cleared. Olivia Beavers and coauthors wrote that the "intensely partisan responses underscore the reality that while Mueller's work is finished, his investigation will loom over the remainder of the current Congress and Trump's presidency" (Beavers *et al.*, 2019).

Some Democrats called for Trump's impeachment as soon as he won the 2016 election. The Democrat Party took control of the House of Representatives in January 2017 and elected Nancy Pelosi to serve as Speaker. The Democrats looked for a reason to impeach Trump by launching several investigations, but the results of their investigations failed to persuade Pelosi and other lawmakers that they could impeach Trump.

A whistleblower complaint about a July 2019 phone call between Trump and Ukrainian President Volodymyr Zelenskyy was made public in September 2019. The acting ambassador to Ukraine corroborated the complaint, which claimed that Trump told Zelenskyy to investigate possible irregularities in the financial dealings of Hunter Biden, son of former Vice President Joe Biden. The complaint alleged that Trump had threatened to withhold U.S. foreign aid funding.

By late November 2019, Democrats were prepared to proceed with the impeachment process because they were sufficiently convinced that they had a case for two articles of impeachment: wrongdoing, and obstruction of Congress. The White House denied the allegations. On December 18, 2019, the House of Representatives voted to impeach President Donald Trump for abuse of power and obstruction of Congress. There were 230 votes in favor of impeachment, 197 against, and 1 present, mostly along party lines.

Following the House's approval of both articles of impeachment, the Senate trial began on January 16, 2020. Chief Justice John Roberts of the U.S. Supreme Court oversaw the trial. On February 5, 2020, the Senate voted to acquit President Trump of both articles in a decision that was essentially along party lines (History.com Trump Impeached, 2021).

14.3.3 *The COVID-19 Pandemic during the Trump Administration*

The first two years of Trump's presidency were hampered by Trump's poor relationship with Democrats, who were pursuing the Russia investigation, and Trump's strained relationships with House Speaker Paul Ryan and Senate Majority Leader Mitch McConnell, both Republicans. The third year of Trump's presidency was much more successful: the United States economy was doing well, unemployment was low, and the world was relatively peaceful. Trump held well-attended rallies during his 2020 presidential campaign for re-election, and many analysts predicted that Trump was heading toward an impressive Electoral College win. The emergence of COVID-19 in late 2019 and early 2020 changed the course of history.

Donald Trump relied on Federal government public health officials, especially American physician and immunologist Anthony Fauci, to guide his administration. Fauci served as director of the National Institute of Allergy and Infectious Diseases (NIAID) from 1984 to 2022. He was chief medical advisor to President Biden from 2021 to 2022. In hindsight, Trump was not well-served by his advisors.

Paul E. Alexander, an epidemiologist, was a Senior Advisor to President Trump for COVID-19 Pandemic policy. Alexander and author Kent Heckenlively wrote that officials in the CDC, the National Institutes of Health (NIH), and the WHO colluded to overthrow Trump by advising the president to enact deleterious responses to the COVID-19 crisis (Alexander and Heckenlively, 2023). Alexander and Heckenlively used documents and emails to support their conclusions. They argued that public health officials, including Anthony Fauci, dismissed elective treatments such as hydroxychloroquine, ivermectin, and vitamin D, supported unnecessary lockdowns and school closures, and imposed questionable mask mandates.

Kentucky Senator Rand Paul, a physician, questioned and then challenged Anthony Fauci's COVID-19 Pandemic recommendations. Paul recognized that Fauci's recommendations were having a significant effect on American society. Furthermore, Paul believed that Fauci was attempting to shut down scientific dissent and was misleading the public about the origins of COVID-19.

Paul wrote a book to document his evidence and indict the failures of public health officials during the pandemic (Paul, 2023). According to

Paul, the origin of the COVID-19 virus was probably the result of gain-of-function research at the Wuhan Institute of Virology in China. Gain-of-function research refers to research designed to help an organism acquire a new ability or property, such as helping a virus become transmissible to humans or from human to human. Gain-of-function can occur through natural selection or laboratory experiments.

Paul believed gain-of-function research in Wuhan was partially funded by the United States government without proper authorization. He believed that Fauci and his scientific allies knew the origin of COVID-19 from the beginning of the pandemic. He thought they could be trying to cover up their role in the origin of COVID-19 and the possibility that their positions made it possible for them to earn consulting fees and royalties from private companies. The cover-up included disparaging anyone who challenged their pandemic narrative. The reputations of many renowned scientists were harmed, and some chose to fight back. Scientists who responded to the attacks on their reputations included epidemiologist Paul E. Alexander (Alexander and Heckenlively, 2023), cardiologist Peter A. McCullough (Leake and McCullough, 2023), and physician and biochemist Robert W. Malone (Malone, 2023).

An early treatment protocol using generic, repurposed medications and supplements was developed by McCullough and his colleagues. They believed their early treatment protocol helped many COVID-19 patients avoid hospitalization and death, but public health officials did not welcome their protocol.

McCullough and his colleagues saw that their encouraging results were at first dismissed and later challenged. The scientific validity of the challenges was called into question when the repurposed drugs used in the early treatment protocol were improperly criticized in professional journals. In addition, the developers of the early treatment protocol were being subjected to censorship and attacks in the media. It appeared that the early treatment protocol was being suppressed and its developers denigrated. Meanwhile, public health officials and most of the media were supporting a COVID-19 treatment that was funded by the United States government and developed by large multinational pharmaceutical companies known as Big Pharma. The new and unproven generation of vaccines was based on mRNA technology and received end use authorization (EUA) from government regulators. Lockdowns were relaxed or lifted when the mRNA vaccines were deployed.

The behavior of government officials and the private sector was reminiscent of President Dwight D. Eisenhower's 1961 farewell address. Eisenhower encouraged Americans to take precautions to prevent the military-industrial complex from acquiring so much influence that it could abuse power. McCullough and his colleagues had similar concerns about public health agencies and Big Pharma (Leake and McCullough, 2023).

We saw in the COVID-19 timeline that Robert W. Malone was an inventor of the original mRNA vaccine technology in the 1980s. He realized during the COVID-19 Pandemic that he was being vilified by most of the media for asking questions and expressing concerns about the efficacy of using drugs based on mRNA vaccine technology. He was also censored by Big Tech, that is, multinational companies that provided social media platforms. Billions of people were taking mRNA-based vaccines without knowing the risks. The Vaccine Adverse Event Reporting System (VAERS) is a repository for reports of many adverse side effects associated with mRNA-based vaccines. Side effects can be relatively mild and go away in a few days. They include headaches, fatigue, and soreness at the injection site. More serious side effects include anaphylaxis, a severe allergic reaction to COVID-19 ingredients; inflammation of the heart muscle (myocarditis); and inflammation of the lining outside the heart (pericarditis).

Malone believed that governments were weaponizing fear of COVID-19 to control public behavior. According to Malone, globalists in such organizations as the World Economic Forum (WEF) and the Bill and Linda Gates Foundation worked with Big Pharma and government regulatory agencies to control the worldwide response to COVID-19. Malone documented his views in *Lies My Gov't Told Me* (Malone, 2023).

14.3.4 *The COVID-19 Pandemic and the 2020 Presidential Election*

The COVID-19 timeline presented above highlights key events during the early days of the COVID-19 Pandemic. The virus first appeared in 2019 and led to an outbreak that had to be confronted during the 2020 election year. Republican incumbent Donald Trump was running against Democrat challenger Joe Biden in a contentious presidential race. Political protests and demonstrations sometimes became riots. Rioters tended to ignore government healthcare recommendations. In addition, people were required to remain in one location during lockdowns, which contributed to a difficult mental health environment.

Left-wing activists joined with wealthy individuals and business leaders in an informal alliance to oppose Trump's reelection. The alliance was "a well-funded cabal of powerful people, ranging across industries and ideologies, working together behind the scenes to influence perceptions, change rules and laws, steer media coverage and control the flow of information" (Ball, 2021). Members of the alliance wanted the public to understand that democracy in America was fragile. They believed their efforts were strengthening the election rather than manipulating it.

The Alliance planned to restructure the American election infrastructure during a pandemic. Table 14.4 lists many of the Alliance's activities.

Molly Ball listed five steps that were needed to achieve an election victory: win the election; win the vote count; win certification of election results, win the Electoral College, and win the transition from the Trump Administration to the Biden Administration (Ball, 2021). The first and most important step was to win the election. Items 1–3 and 5–8 helped win the election. Items 1, 2, 4, 8, and 9 helped win the vote count.

Item 2 includes the collection of tens of millions of dollars from a variety of foundations to fund election administration. For example, Mollie Hemingway reported that $400 million was contributed by the

Table 14.4 Activities of the Alliance (Ball, 2021)

Item No.	As reported by Ball, the Alliance …
1	Got states to change voting systems and laws
2	Helped secure hundreds of millions of dollars in public and private funding
3	Fended off voter-suppression lawsuits
4	Recruited armies of poll workers
5	Got millions of people to vote by mail for the first time
6	Successfully pressured social media companies to take a harder line against disinformation
7	Used data-driven strategies to fight viral smears
8	Executed national public-awareness campaigns
9	Monitored every pressure point after the election

Chan Zuckerberg Initiative (Hemingway, 2021a) and discussed ostensibly legal steps that were taken to "rig the election" (Hemingway, 2021b). Money was used to effectively privatize election administration in swing cities and states where a relatively small shift of voters toward Democrat candidates could alter a close election. Ball acknowledged that most analysts expected "a 'blue shift' in key battlegrounds — the surge of votes breaking toward Democrats, driven by tallies of mail-in ballots — but they hadn't comprehended how much better Trump was likely to do on Election Day."

Public awareness campaigns in item 8 of Table 14.4 told Americans to expect the vote count to last for days or weeks beyond Election Day in November 2020. The monitoring in item 9 was designed to prevent the Trump campaign from overturning the election.

After the November 2020 election, Donald Trump told his supporters on January 6, 2021 that lawmakers or Vice President Mike Pence could reject states' electoral votes. Trump was relying on the Electoral Count Act of 1887 which gave Congress rules for handling electoral votes (Parks, January 2022). Trump's opponents and some of his allies claimed that Trump had no legal basis to reject electoral votes. Despite their claim, Democrat-controlled Congress reformed the Electoral College Act later in 2022 (Parks, December 2022).

Trump picked former New York City Mayor and U.S. Attorney Rudy Giuliani to head the legal efforts to challenge the results after Trump lost the 2020 presidency to Joe Biden. Giuliani failed to overturn the election results and his efforts led to several lawsuits (BOE Giuliani, 2023). Item 9 in Table 14.4 shows that the Alliance anticipated legal challenges and was prepared to defend Biden's victory.

Mollie Hemingway analyzed the 2020 election and wrote a definitive account of irregularities that tended to favor Biden (Hemingway, 2021b). For example, Big Tech companies minimized the reach of Trump supporters in social media, the 21st century public square. Credible accounts about Biden family corruption were labeled misinformation and essentially censored before the election. Stories of mail-in ballot abuse were common. Voting rolls were used to provide addresses for mail-in ballots even though they were not up-to-date and included addresses of voters who died (ghost voters) or moved. Democrat operatives manipulated the voting process during the COVID-19 Pandemic. Americans, especially Trump supporters, questioned the fairness of an election that saw Trump receive more votes than any other Republican presidential candidate had ever received on Election Day and then watched his lead evaporate as mail-in ballots were received and counted several days after Election Day.

14.3.5 Was the 2020 Presidential Election a Coup?

Some authors have suggested that the 2020 Presidential Election was a coup. Patrick Colbeck presented evidence for the coup in his book entitled *The 2020 Coup* (Colbeck, 2022). Attorney Christina Bobb made the observation that Democrat Party officials were corrupt during the 2020 election, and Republican Party officials chose to remain silent or ignore election irregularities during the 2020 pandemic year (Bobb, 2023).

In his assessment of the 2020 election and its implications for the 2024 election, David Horowitz referred to the mass migration into the United States during the Biden administration and said that "Barbarian terrorist forces are already at the gates [of the United States], and inside them. American leaders—both military and civilian—are preoccupied with delusional threats that are said to be existential—climate change, white supremacy, patriotic extremism. But the greatest existential threat to American democracy is the drive by the Democrat Party to create a one-party socialist state—a fascist state" (Horowitz, 2022, p. 211). He cautioned that "empires and states rise and fall while everybody is watching. Although the watchers may be surprised when the actual collapse occurs, with the hindsight provided by the end itself, everybody can see how it fell" (Horowitz, 2022, p. 211).

Mark Levin was a top adviser to several members of Republican Ronald Reagan's Presidential cabinet. Levin discussed the leftward movement of the Democrat Party in *American Marxism* (Levin, 2021). Two years later Levin pointed out in *The Democrat Party Hates America* that "the Democrat Party is more than a political party. It is the state party. It seeks to monopolize the political system, the culture, government, and society" (Levin, 2023, p. 327). To achieve its fundamental transformation of the United States, Levin said that "The Democrat Party uses the culture and politics to empower itself and its agenda … when the Democrat Party wins elections, it claims broad mandates; when it loses elections, it ignores the popular will of the people and turns to the permanent government and its cultural surrogates to sabotage the Republicans and push forward their American Marxist agenda" (Levin, 2023, p. 328). Levin concluded that "every legal, legitimate, and appropriate tool and method must be employed in the short- and long-run to shatter the Democrat Party... The Democrat Party must be resoundingly conquered in the next election and several elections thereafter, or it will become extremely difficult to undo the damage it is unleashing at breakneck pace" (Levin, 2023, p. 330).

14.3.6 *The 2021 Donald Trump Impeachment*

On January 13, 2021, President Trump was impeached by the House of Representatives, once again under the leadership of Speaker Nancy Pelosi. Trump was impeached for inciting the January 6, 2021 riot at the U.S. Capitol. He was the first U.S. president to be impeached twice. The Democrats and ten House Republicans voted in favor of impeachment. Donald Trump was replaced as president on January 20, 2021, when Joe Biden was inaugurated. The second impeachment trial of Donald Trump began on February 9, 2021. Seven Republican senators voted with Democrat senators to convict Trump, but they did not have the 60 votes needed to convict. The trial ended on February 13, 2021, with Trump's acquittal.

14.3.7 *The International Response to the COVID-19 Pandemic*

The above discussion focused on the impact of civil rights abuses in the United States. These abuses occurred worldwide. In this section, we illustrate civil rights abuses in a few international examples.

China

The COVID-19 virus was first detected in Wuhan, China in 2019. Xi Jinping, the leader of China, tried to stop the COVID-19 outbreak by imposing a zero-COVID policy. The goal of Xi's zero-COVID policy was to keep the number of cases as close as possible to zero. To achieve this goal, China enforced strict lockdowns, quarantines, and mass testing.

Enforcement of Xi's policies helped minimize the number of COVID-19 cases, but significantly slowed the economy and fueled public discontent. A surprisingly large number of COVID-19 infections spread through China. In 2022, a strict lockdown was imposed April 1 on Shanghai, a city of 25 million people and a financial hub. The lockdown continued for two months and impacted the global supply chain as Shanghai manufacturers were unable to operate. When the lockdown ended, there was "an outpouring of relief, joy and some wariness from exhausted residents" (Goh and Woo, 2022).

New Zealand

Labour Party candidate Jacinda Ardern became the youngest Prime Minister of New Zealand in October 2017 at age 37. She was Prime Minister on February 2, 2020, when a man in the Philippines became the first person outside of China to die of COVID-19. New Zealand banned entry to any foreigners traveling through or from China even though there were no known cases in New Zealand. If New Zealand residents returned from China, they were isolated for 14 days.

Ardern continued to tighten travel bans. By March 2020, she had banned almost every resident or foreigner from entering New Zealand. In addition, she imposed "a strict nationwide lockdown when just over 100 cases of COVID-19 … had been confirmed" (BOE Ardern, 2023).

Ardern's policies damaged New Zealand's economy, which had a significant dependence on tourism, but the policies were effective. By closing the border early in the pandemic, the number of COVID-19 cases was minimized. Ardern won reelection in 2020. She was soon confronting waves of Delta and Omicron variants of COVID-19. Ardern's government "shifted from lockdowns to an emphasis on vaccination and acceptance of 'living with the virus'" (BOE Ardern, 2023).

Opposition to Ardern's policies continued to grow. Protests of lengthy lockdowns were followed by opposition to vaccination requirements. In addition, economic and social issues were increasingly worrisome and seemed to be a low priority to a government focused on the pandemic. The decline in Ardern's popularity and the toll the pandemic had on her led her to announce on January 9, 2023, that she would "stand down as prime minister by February 7" (BOE Ardern, 2023). The Labour Party chose Chris Hipkins as her successor. Hipkins became Prime Minister of New Zealand on January 25, 2023.

Australia

Three philanthropic trusts commissioned a study of Australia's response to the COVID-19 Pandemic. The study was conducted by a panel of four people including its leader Chancellor Peter Shergold of Western Sydney University. Shergold served as head of the Department of Prime Minister and Cabinet during former Australian Prime Minister John Howard's administration. The panel published a report entitled Fault Lines on October 20, 2022 (Shergold *et al.*, 2022). The study mainly considered the period when Scott Morrison was Prime Minister of Australia. Morrison served from 2018 to 2022 and was succeeded by Anthony Albanese in May 2022.

The Shergold report was based on a six-month study that included over 350 confidential submissions and consultations with health experts, public servants, economists, and business and community groups. According to Chip Le Grand of *The Sydney Morning Herald*, the Shergold report "found ill-conceived policies, politically driven health orders and excessive use of lockdowns failed to protect the old, disregarded the young and abandoned some of the nation's most disadvantaged communities" (Le Grand, 2022). Table 14.5 shows areas where the response could have been improved.

Le Grand highlighted a list of decisions that were considered important but wrong. The list is shown in Table 14.6. JobKeeper in Item 1 was a government-funded program that provided biweekly payments to eligible Australians for six months. Triple jeopardy in Item 4 refers to individuals with disabilities who were more likely to become critically ill from the virus and could lose access to regular health care, prescription

Table 14.5 What Could Have Been Done Better? (Shergold *et al.*, 2022)

Item No.	Comment
1	Economic supports such as sick leave and supplemental income should have been provided fairly and equitably.
2	Lockdowns and border closures should have been used less.
3	Schools should have stayed open.
4	Older Australians should have been better protected.

Table 14.6 Consequential Incorrect Decisions (Le Grand, 2022)

Item No.	Comment
1	Casual workers and temporary migrants were excluded from JobKeeper and a clawback mechanism was needed to retrieve payments from highly profitable companies.
2	Schools were shutdown without adequate consideration of the long-term effects on learning, social development, mental and physical health, and the national economy.
3	State health officials were reluctant to hospitalize elderly care residents who tested positive for COVID.
4	Inadequate support for people facing "triple jeopardy."

drugs, therapeutic services, and social interaction with family, friends, and coworkers.

Canada

Canadian truckers began "Freedom Convoy" protests to oppose a Canadian government requirement that cross-border drivers must be quarantined or vaccinated against COVID-19. Copycat demonstrations appeared in Australia, France, Israel, and New Zealand. Canadian Prime Minister Justin Trudeau declared a state of emergency on February 14, 2022, to end the protests that had paralyzed sections of the capital and closed several border crossings with the United States, including the busiest border crossing between the U.S. and Canada. Trudeau believed that "The blockades are harming our economy and endangering public safety" (Reuters Canadian Truckers, 2022).

The Canadian Emergencies Act authorized the government to stop funding to protesters and strengthened local and provincial law enforcement by providing federal police support. Financial measures of the Emergencies Act enabled Canadian banks to freeze accounts suspected of funding blockades, suspension of insurance on vehicles involved in protests, and oversight of crowdfunding platforms using terror-finance laws. Finance Minister Chrystia Freeland said that "We are making these changes because we know that these (crowdfunding) platforms are being used to support illegal blockades and illegal activity which is damaging the Canadian economy" (Reuters Canadian Truckers, 2022).

The Canadian Civil Liberties Association (CCLA) said that the Emergencies Act was supposed to be used to deal with threats to Canada's sovereignty, security, and territorial integrity. Members of the CCLA did not believe the government had met the standard for invoking the Emergencies Act. Opposition politicians said emergency measures were unnecessary and an abuse of power. Conservative Party Member of Parliament Dean Allison said that the Emergencies Act permitted "authoritarian military-style measures" against the protesters (Reuters Emergency Powers, 2022). Green Party member Mike Morrice said "the use of the Emergencies Act sets a worrying precedent for future protests" (Reuters Emergency Powers, 2022).

Emergency measures had to be approved by the Canadian Parliament within seven days. The Canadian House of Commons supported Trudeau's decision to invoke emergency powers on February 21. The Canadian Senate was debating the act on February 23 when Trudeau revoked the act, saying "The situation is no longer an emergency" (Boisvert, 2022).

Although the use of the Canadian Emergency Act was short-lived, it showed how much power could be used against the Canadian population by the Canadian government.

Point to Ponder: Freedom or Security?

The COVID-19 Pandemic showed the world that fear of a life-threatening event could be used to abuse civil rights by imposing limitations on freedom, such as lockdowns, censorship of free speech, vilification and censure of dissenters, defunding individuals or groups, and manipulating Artificial Intelligence (AI) algorithms. The use of financial control is a growing concern in countries that are seeking to centralize banking and replace cash with digital currency.

14.4 The Great Reset

Klaus Schwab of the WEF wanted to take advantage of the COVID-19 Pandemic to implement the Great Reset (Schwab, 2020). He said that COVID-19 responses in the first half of 2020 demonstrated that the social and economic foundations of society should be reset. The goal of the

Great Reset was to implement stakeholder capitalism, which is a form of capitalism that requires private sector firms to create long-term value by considering the needs of society and all stakeholders.

According to Schwab, the world needs to move quickly and cooperatively to reform every aspect of national economies and societies. The Great Reset of capitalism requires that every nation and industry must be transformed. Schwab argued that COVID-19 provided the most urgent justification for implementing the Great Reset. He said that COVID-19 deaths, increasing government debt, increasing unemployment (due in part to lockdowns), weakening of environmental protections and enforcement, and inequality between the wealth of billionaires and everyone else will exacerbate existing climate and social crises.

The pandemic has shown how quickly we can make drastic lifestyle changes. The crisis compelled people and organizations to change long-held beliefs and practices. For example, frequent air travel was replaced by remote meetings, and commuting to work every day was replaced by remote work and occasional meetings at the office.

The public demonstrated a willingness to make sacrifices so that health care and essential services could be provided to people in need, such as vulnerable groups like the elderly. In a move toward stakeholder capitalism, many businesses took action to support their employees, clients, and local communities.

Schwab looked at human behavior during the pandemic and concluded that society was willing to build a better society by enacting the Great Reset. He said the Great Reset would have three key components. Table 14.7 summarizes Schwab's key components. Schwab concluded that the "pandemic represents a rare but narrow window of opportunity to

Table 14.7 Components of the Great Reset (Schwab, 2020)

Component No.	The Great Reset would…
1	Guide the market toward fairer outcomes.
2	Ensure that investments advance shared goals based on ESG metrics (see Section 13.4).
3	Harness the innovations of the Fourth Industrial Revolution to support the public good (see Section 11.5).

reflect, reimagine, and reset our world to create a healthier, more equitable, and more prosperous future" (Schwab, 2020).

Not everyone was willing to adopt Schwab's Great Reset. Glenn Beck and Justin Haskins wrote that the Great Reset is driven by the world's elite to control the global economy using banks, government programs, and ESG metrics (Beck and Haskins, 2022). If the Great Reset is completed, Beck and Haskins believe that already powerful multinational organizations, such as the United Nations, and many of the companies represented by attendees at the WEF, would acquire additional economic and social power.

In a more recent book, Beck, Haskins, and contributor Donald Kendal wrote that factors such as advances in technology and cultural shifts will result in extreme social disruption (Beck and Haskins, 2023). The authors expect advocates of the Great Reset to take control of economies and societies in North America and Europe by exploiting the rejection of traditional values.

Author Alex Jones is another influential critic of the Great Reset. Jones said in his book *The Great Reset* that "The premise of this book is that the battle we are fighting against the Great Reset is nothing more than an ancient battle between the forces of freedom and tyranny" (Jones, 2022, p. 11). He went on to present an analysis of Schwab's view of the Great Reset and concluded that "We are in a war for the future of the world. The globalists want an anti-human future in which they will capture control of our species and direct the future of human development. But that's not a future the people want. They want a future of freedom, where scientific discoveries liberate them to be so much more than they could ever be when they were focused on mere survival or dealing with the ravages of disease" (Jones, 2022, p. 234).

Author Douglas Murray agreed that a war was being waged for the future of the world, but the war was being waged against the roots of Western Civilization and the benefits that have emerged from Western traditions. In his book *The War on the West*, Murray wrote "about what happens when one side in a cold war—the side of democracy, reason, rights, and universal principles—prematurely surrenders" (Murray, 2022, p. 6). He presented arguments that support the roots, traditions, and accomplishments of Western Civilization.

Murray was defending Western Civilization against a cultural revolution that author Christopher Rufo said was led by the radical left. Rufo described his study of "the ideology that drives the politics of the radical left" (Rufo, 2023, p. x) in *America's Cultural Revolution*. He said that the cultural revolution is attempting to negate "the metaphysics, morality, and stability of the common citizen. As [the cultural revolution] undermines the institutions of family, faith, and community, it creates a void in the human heart that cannot be filled with its one-dimensional ideology" (Rufo, 2023, p. 279).

Historian Victor Davis Hanson expressed concern in 2021 that the American citizen was under attack by progressive elites, tribalism, and globalization in his book *The Dying Citizen*. Hanson pointed out that the years "2017 to 2019 had seen progress in restoring the sanctity of American citizenship, an effort rendered ever more controversial by the support and efforts of Donald Trump" (Hanson, 2021, p. 345) following 8 years of the Obama–Biden administration. Hanson realized that many of the gains made by the middle class during the early Trump years were lost during the 2020 pandemic and presidential election year. He concluded that "As 2021 began, the supporters of restoring the primacy of the American citizen were not so confident in their own powers of renovation as they were convinced that they had no other choice but to keep trying.

"The stakes were no less than the preservation of the American republic itself" (Hanson, 2021, p. 345).

Chapter 15
What Does the Future Look Like?

In Part 1, we provided a succinct review of how the existing world order emerged and introduced several factors that were capable of catalyzing changes to the existing world order. Mechanisms for reshaping the existing world order were discussed in Parts 2–4. We have seen how the environment, climate change, institutions, and events can be weaponized to advance political agendas. In this chapter, we consider a set of scenarios that could result in a new world order after the energy transition. We begin by considering a modern study of the changing world order.

15.1 The Changing World Order

We introduced the study of the rise and decline of empires in Chapter 2. In this section, we consider the work of Ray Dalio, founder and chief investment officer of Bridgewater Associates, a global asset manager. Dalio developed a model of the changing world order based on his study of history (Dalio, 2021). He identified several significant developments that had happened many times throughout history but not during his lifetime. Dalio believed that history transitions from one world order to another through relatively well-defined stages.

The empires and dynasties studied by Dalio rose and fell in a classic Big Cycle with markers that let us see where we are in the Big Cycle. The Big Cycle oscillates between two extremes. The first extreme is characterized by the significant improvement of living standards resulting from times of creativity and productivity that were peaceful and prosperous. The second extreme is characterized by periods of depression, revolution, and war marked by vigorous power struggles and widespread destruction of life, wealth, and other valuables.

Dalio observed that "no system of government, no economic system, no currency, and no empire lasts forever, yet almost everyone is surprised and ruined when they fail" (Dalio, 2021, p. 5). He explained how a new world order can be created with a simple narrative.

According to Dalio, people have historically fought over many issues such as ideology and religion. However, the struggle to create, seize, and divide power and money has been the primary challenge facing most people worldwide. The wealthiest people tend to be those who control the means of producing wealth. Wealthy people typically foster mutually beneficial relationships with people in positions of political power. The wealthy and politically powerful worked together to establish and enforce rules that helped them preserve or grow their wealth and power.

A very small percentage of the population can gain and control extraordinarily large percentages of wealth and power. Dalio listed the determinants of wealth and power as "1) education, 2) competitiveness, 3) innovation and technology, 4) economic output, 5) share of world trade, 6) military strength, 7) financial center strength, and 8) reserve currency status" (Dalio, 2021, p. 41). Reserve currency is a currency that is accepted worldwide for transactions and savings. It is the most essential component of international capital markets and economies (Dalio, 2021, p. 12). The US dollar is the current reserve currency, but it is being challenged by BRICS (see Section 13.5).

Over time, the wealthy and powerful encounter difficulties that hurt the poor and weak in society. The gap between the "haves" and "have nots" of society can lead to conflicts and, at times, revolutions, or civil wars. Conflict can cause a new world order to emerge, and a new Big Cycle begins.

Dalio's study showed that "great empires typically lasted roughly 250 years, give or take 150 years, with big economic, debt, and political cycles within them lasting about 50 to 100 years." The economic, debt, and political cycles that appear to govern the Big Cycle are shown in Table 15.1.

Figure 15.1 shows an archetypical Big Cycle with three periods: the rise, the top, and the decline (Dalio, 2021, p. 43). The Big Cycle represents the transition from one world order to another.

The advance, or evolution, of society can be depicted as a corkscrew advance of Big Cycles, as shown in Figure 15.2 (Dalio, 2021, p. 28).

Table 15.1 Cycles that Influence the Rise and Decline of Empires (Dalio, 2021)

Item	Cycle
1	The long-term debt and capital markets cycle
2	The internal order and disorder cycle
3	The external order and disorder cycle

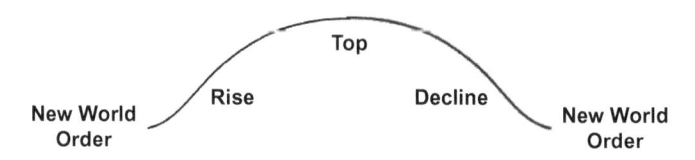

Figure 15.1 The Archetypical Big Cycle

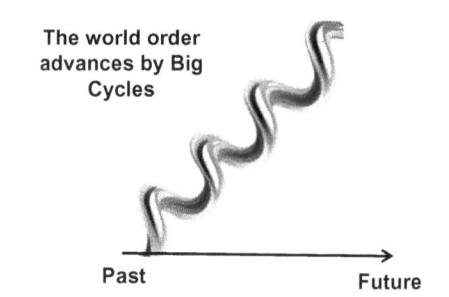

Figure 15.2 Corkscrew Evolution of the World Order

Dalio imagined the evolution of Big Cycles as an ascending trajectory toward societal advancement brought about by adaptation and learning. He considered human productivity the primary driver of the global increase in wealth, power, and living standards over time. However, Dalio pointed out that "learnings and productivity improvements are evolutionary; they don't cause big abrupt shifts in who has what wealth and power. The big, abrupt shifts come from booms, busts, revolutions, and wars, which are primarily driven by cycles, and these cycles are driven by logical cause/effect relationships" (Dalio, 2021, p. 31).

Table 15.1 specifies the cycles that drive the Big Cycle. Table 15.2 lists six stages of the long-term debt cycle. Debt is related to money, which is a medium of exchange that can also store wealth. In stage 1,

Table 15.2 Six Stages of the Long-Term Debt Cycle (Dalio, 2021, pp. 116–128)

Stage	Comment
1	The cycle begins with (a) little or no debt and (b) money being "hard"
2	Then comes claims on hard money
3	Then comes increased debt because credit is allowed
4	Then debt crises, defaults, and devaluations come, which leads to the printing of money and the breaking of the link to hard money
5	Then comes fiat money, which eventually leads to the debasement of money
6	Then the flight back into hard money

financial transactions occur with the exchange of hard currency such as precious metals. In stage 2, notes or paper money are allowed to serve as claims on hard currency. Stages 3 and 4 signal the introduction of credit and the devaluation of money. By stage 5, money is no longer tied to hard currency. The fiat monetary system allows the printing of paper money by a central bank. By stage 6, people are trying to preserve their wealth.

The second cycle in Table 15.1 is the Internal Order and Disorder Cycle. This cycle has six stages as shown in Table 15.3. The Internal Order and Disorder Cycle ends in internal conflict.

The third cycle in Table 15.1 is the External Order and Disorder Cycle. This cycle can follow the six stages shown in Table 15.3 with one key exception: "international relations are driven much more by raw power dynamics" (Dalio, 2021, p. 194). Dalio recognized that "attempts have been made to make the external order more rule-abiding (e.g., via the League of Nations and the United Nations), by and large they have failed because these organizations have not had more wealth and power than the most powerful countries" (Dalio, 2021, pp. 194–195). Table 15.4 lists the major types of fights that occur when the External Order and Disorder Cycle ends in external conflict, that is, conflict or war with another state.

Table 15.3 Six Stages of The Internal Order and Disorder Cycle (Dalio, 2021, p. 152)

Stage	Comment
1	The new order begins and the new leadership consolidates power
2	Resource-allocation systems and government bureaucracies are built and refined
3	A period of peace and prosperity ensues
4	A period with great excesses in spending and debt is accompanied by the widening of wealth and political gaps
5	A period with very bad financial conditions and intense conflict
6	A period when civil wars/revolutions occur

Table 15.4 Five Major Types of Conflicts between Nations (Dalio, 2021, p. 194)

Item	Type of Conflict	Typical Features
1	Trade/economic wars	Tariffs, import/export restrictions
2	Technology wars	Access to and control of technology
3	Geopolitical wars	Disputes over territory and alliances
4	Capital wars	Use of sanctions and access to capital
5	Military wars	Use military forces to settle disputes

Dalio has applied his Big Cycle model of geopolitics to the modern world and predicted that the United States is in decline as leader of the current world order. China, on the other hand, is on the rise as the leader of the next world order (Dalio, 2021, pp. 150–151).

15.2 Scenarios

A scenario describes a hypothetical set of actions or events that could occur in the future. The sampling of scenarios presented here illustrates what the world could become if the trends identified in previous chapters continue.

15.2.1 *Scenario 1: Globalization*

The first scenario is similar to Samuel P. Huntington's One Unified World paradigm discussed in Chapter 2. A supranational authority controlled by a ruling class seems to be the destination of many of the trends discussed in Part IV. The United Nations is the current contender for becoming the supranational authority.

According to Scenario 1, the world of tomorrow will be governed by a supranational government that needs substantial enforcement power. The weaponization of the COVID-19 Pandemic by central governments showed us that the ruling class can call upon physical control of the public as well as control by technological tools such as surveillance and central control of digital currency.

15.2.2 *Scenario 2: A Bipolar World*

The second scenario is similar to Samuel P. Huntington's Two Worlds paradigm discussed in Chapter 2. The world is polarized into two geopolitical coalitions. As noted in Chapter 2, nations are presently aligning with either democracies, or totalitarian regimes such as China and Russia. The coalition of democracies is called here the Western Alliance, and the coalition of totalitarian regimes is called the Eastern Axis.

The Western Alliance is led by the United States and NATO, while the Eastern Axis is led by China, Russia, and Iran. The Eastern Axis leaders are also members of BRICS, discussed in Chapter 13, and China is leading the Belt and Road Initiative (BRI). BRICS is mounting a challenge to the US$, which is the current global currency.

Harvard University scholar Graham Allison studied 16 cases of encounters between an existing power and a rising power over the past 500 years. Allison and his colleagues found that the most likely outcome of the encounter is armed conflict. In 12 of the 16 cases the encounter led to war. Allison called the encounter the "Thucydides Trap" (Allison, 2015; Mosher, 2017, p. 34). Thucydides wrote the *History of the Peloponnesian War* between Athens and Sparta in the 5th century B.C. Thucydides said that armed conflict was the result of a rapid shift in the balance of power between the established power and the challenging power.

Allison's results suggest that the most likely outcome of an encounter between the Western Alliance and the Eastern Axis is armed conflict.

The conflict could be started by one or more factors, such as a clash of cultures, and competition for valuable resources. Disputed resources include fossil fuels, fissile materials, and electric vehicle (EV) materials such as lithium.

Developed nations want access to natural resources that are widely distributed geographically. Nations that are not aligned with either the Western Alliance or the Eastern Axis are concerned that their resources could be seized. This suggests that a third scenario is possible.

15.2.3 *Scenario 3: A Tripolar World*

The bipolar world in Scenario 2 is expanded to a tripolar world in Scenario 3. Nations that do not align with the Western Alliance or the Eastern Axis establish a third coalition of non-aligned states. U Thant showed that non-aligned states can influence global events using a supranational authority like the United Nations.

15.2.4 *Scenario 4: A Multipolar World*

The fourth scenario is similar to Samuel P. Huntington's Anarchy discussed in Chapter 2. In this scenario, nations pursue their own interests but may be loosely bound by common interests to create a multipolar world with multiple power centers.

Daniel Yergin said that the world "has become more fractured, with a resurgence of nationalism and populism and distrust, great power competition, and with a rising politics of suspicion and resentment" (Yergin, 2020, p. 423). Support for globalization would remain but become more fragmented and confrontational. Yergin realized that several significant obstacles stand in the way of a rapid transition to sustainable energy worldwide. Key obstacles include "the sheer scale of the energy system that supports the world economy, the need for reliability, the demand for mineral resources for renewables, and the disruptions and conflicts that would result from speed" (Yergin, 2020, p. 428).

In his book *Global Public Governance: Toward World Government?* Sorpong Peou examined global public governance within the context of eight key topics: global security, human rights, global criminal justice, global health, global education, global finance, global trade, and the global environment (Peou, 2022). Peou concluded that multiple regions could provide a more pluralist world order in which distinct ethnic, religious, or cultural groups could have more representation. A regional approach to public governance could be more effective than globalization and global public governance.

15.2.5 Scenario 5: Chaos

The fifth scenario is similar to Samuel P. Huntington's Chaos discussed in Chapter 2. In this scenario, the importance of nation states fades as people form loose alliances based on common traditions and value systems. These alliances are facilitated by global mass communication and the availability of information through technology such as social media.

15.2.6 Scenario 6: Global Feudalism

Proponents of anthropogenic climate change believe that globalization and a sense of urgency are needed to meet environmental challenges. Globalists tend to favor open borders and multinational agreements. The British exit from the European Union (BREXIT) and the 2016 election of Donald Trump as President of the United States show voter concern on both sides of the Atlantic about the value of globalism and the loss of national sovereignty, including loss of control of foreign policy and immigration policy.

Supranational agreements designed to meet the challenges associated with anthropogenic climate change are favored by advocates of central planning on a global scale. In *The Road to Serfdom*, Austrian economist and philosopher Hayek (1944) warned that tyranny could result from government-controlled central planning. Hayek observed that every step away from the free market and toward government planning is a step toward more government control in exchange for a loss of human freedom.

The emergence of a global ruling class that rejects fossil fuels in favor of an urgent transition to renewable energy requires the growth of a global centralized authority that can enforce the transition. Individuals will lose freedoms and could become part of a global feudal society consisting of a global ruling class and everyone else.

Joel Kotkin described the emergence of a form of feudalism in America (Kotkin, 2019), especially in California (Kotkin, 2020). Two classes exercised power in feudal society: the clergy, and the warrior-aristocracy. The French referred to the clergy class as the First Estate, and the warrior-aristocracy class as the Second Estate. All others, including prosperous merchants, were members of the Third Estate. Most of the Third Estate consisted of subsistence-level peasants. Today, society in many Western democracies is tending toward a type of feudalism based on declining material progress and social mobility, and great concentration of wealth among a small group (Kotkin, 2019). In California, Kotkin describes a "feudalism characterized by gross inequality and increasingly rigid class lines" Kotkin, 2020). The decline of middle-classes presages the rise of feudalism.

The Goldilocks Policy for energy transition minimizes the risk of enabling a totalitarian state such as global feudalism by establishing a framework that can be adopted worldwide. There is time to develop and adopt reliable and cost-effective energy sources as the world makes the transition to a sustainable energy mix.

15.3 Concluding Remarks

We have presented the historical basis of modern political movements that are influencing the energy transition and driving a change in global governance. For example, some globalists are seeking to change the existing world order by establishing a supranational authority. Another group consists of totalitarians who want to replace the existing world order, which allows individual freedom, with a world order based on state control of individuals. Advocates for many of the movements we have discussed have been working for decades to achieve their objectives.

We must understand that new technologies are being used to control social behavior and recognize the forces that are driving global trends. Several authors have proposed strategies for countering government expansion and control of the citizenry (for example, Gonzalez and Gorka, 2022; Cruz, 2023; Jones, 2023; Levin, 2023; Rufo, 2023). These strategies require citizens to take action in the public square, social media, and the ballot box to help shape the future.

People who embrace freedom often remind us to remember American founder Benjamin Franklin's admonishment that "Those who would give up essential Liberty, to purchase a little temporary Safety, deserve neither Liberty nor Safety" (Franklin, 1755). Freedom lovers still have time to influence the outcome of events.

Appendix

The Goldilocks Policy for Energy Transition

A.1 Introduction

The Goldilocks Policy for energy transition demonstrates how past energy transitions can be used to estimate a suitable time frame for an orderly transition to a sustainable energy portfolio. An updated forecast of global energy consumption that includes the COVID pandemic period and global energy consumption data through 2022 is presented here.

A.2 How Much Time Do We Have?

An examination of energy consumption as a function of time and the duration of energy transition periods in the United States provides an estimate of the time needed to achieve an energy transition.

Figure A.1 shows the contribution of different energy sources to the United States energy mix during the period from 1775 to 2022. The energy category labeled "Nuclear Electric" refers to electricity generation by nuclear fission reactors. The energy category labeled "Other Renewable" includes wood, geothermal, solar thermal, photovoltaic, and wind.

The data in Figure A.1 can be rearranged to approximate the length of time it has taken historically for a developed nation to transition from one energy source to another. Figure A.2 displays the percent contribution of different energy sources to the United States energy mix during the period from 1775 to 2022.

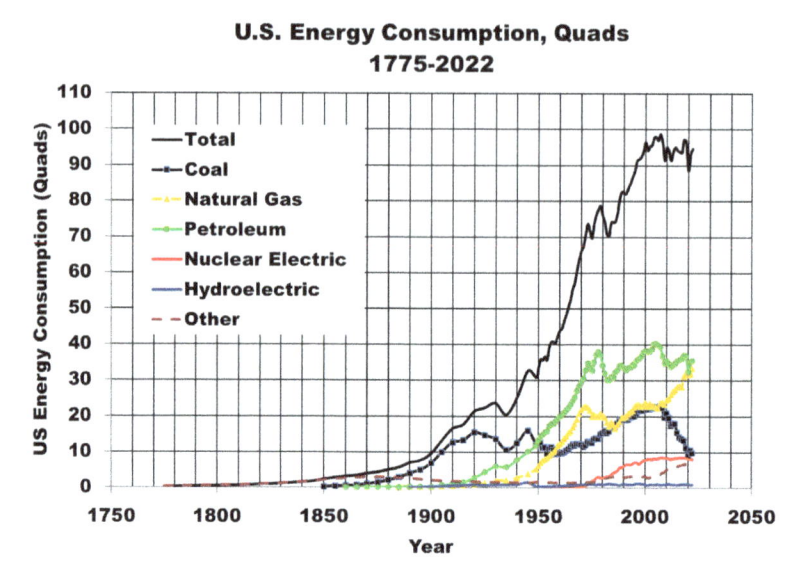

Figure A.1 US Energy Consumption by Source, 1775–2022 (1650–1945 data from US EIA AER, 2011; remaining data from US EIA MER Table 1.3, September 2023)

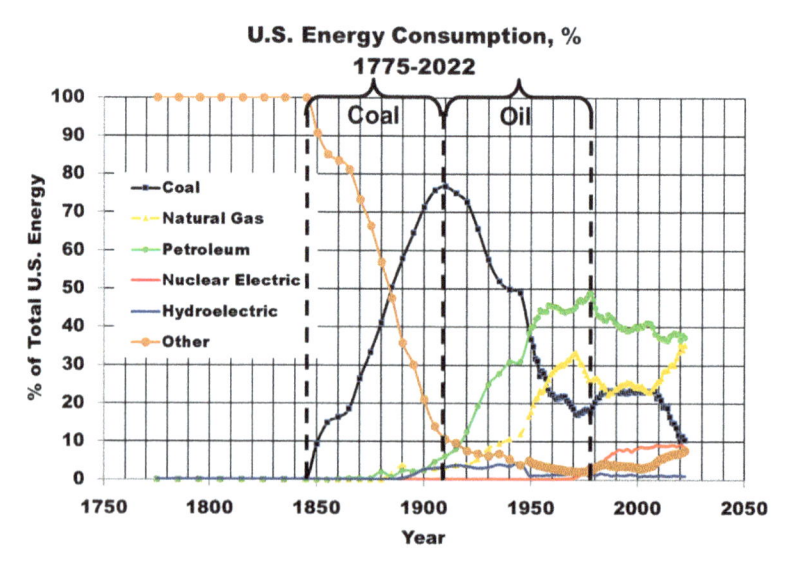

Figure A.2 Coal and Oil Transition Periods based on US Energy Consumption by Source, 1775–2022 (%)

Estimates of coal and oil transition periods in the United States are shown in the figure. The transition period begins when the leading energy source begins to decline. In Figure A.2, wood is the first leading energy source. The consumption of wood begins to decline when another energy source, coal, is available and adopted for use in the mid-1800s. Petroleum began to replace coal in the early 1900s and peaked in the latter half of the 20th century. It took approximately 60–70 years to transition from wood to coal and then from coal to oil.

A.3 The Goldilocks Policy and the 2% Solution

Fanchi and Fanchi (2015) introduced the Goldilocks Policy for Energy Transition as a policy that would implement a transition to a sustainable energy mix based on historical energy transition periods. It is called the Goldilocks Policy because it relies on an energy transition period that is neither too fast nor too slow, but just right; that is, the Goldilocks Policy should be based on an implementation plan that minimizes environmental impact and reduces uncertainty in business planning with predictable public policy. If the Goldilocks Policy is not applied, government policy may vacillate between an energy transition that is so fast that it could significantly damage the global economy, or an energy transition that is so slow that it could permanently damage the environment.

One way to implement the Goldilocks Policy is to adopt an energy transition rate of 2% per year. This energy transition rate is called the Two Percent Solution. It fits within the historically observed energy transition rate discussed previously. According to Lindsey and Dahlman (2022), the average global temperature is predicted to rise by 0.36°F (0.22°C) per decade. The average global temperature would rise about 2.16°F (1.32°C) if the temperature increases at an average rate of approximately 0.36°F (0.22°C) per decade during an energy transition period of 60 years.

We can use the Goldilocks Policy to forecast energy consumption given a few assumptions. In this calculation, we assume that energy consumption will continue the linear growth it has shown in this century, and that the consumption of nuclear fission energy will not change. The Goldilocks Policy calls for increasing the consumption of alternative energy by 2% per year to match the decline in fossil fuel consumption. The energy consumption forecast shown in Figure A.3 includes the COVID pandemic period (CDC COVID-19, 2022). The forecast in

[Chart: 2% Solution of the Goldilocks Policy — Energy Consumption (Quads) vs Year, showing Total, Fossil Fuels, Nuclear, and Other from 1980 to 2100]

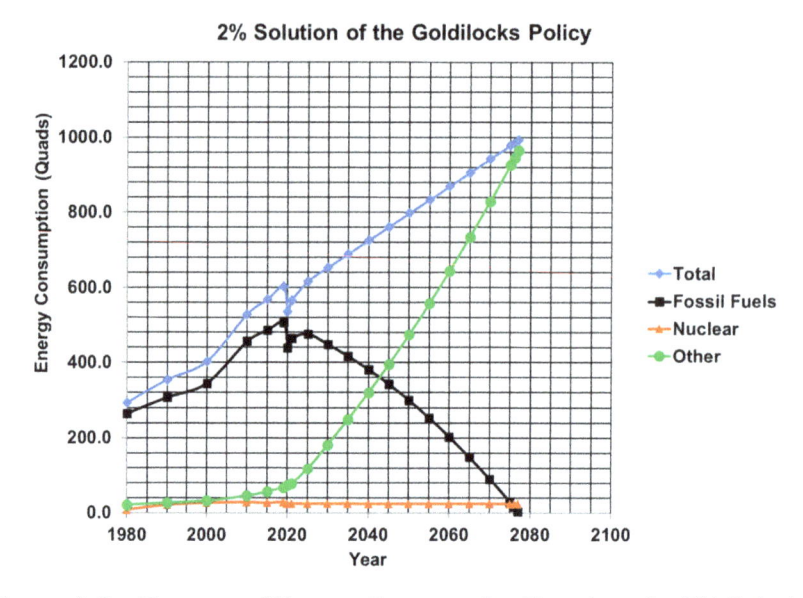

Figure A.3 Forecast of Energy Consumption Based on the 2% Solution of the Goldilocks Policy (Updated from Fanchi, 2024)

Figure A.3 shows that fossil fuel consumption will end by 2080 unless society decides to continue some reliance on fossil fuels.

A.4 The Future

The Goldilocks Policy is based on a vision that acknowledges the need to safeguard the environment from fossil fuel combustion while preserving regional, national, and international economies. Implementation of the Goldilocks Policy faces several challenges. Many of the challenges are discussed in more detail elsewhere (Fanchi, 2019). Additional challenges have appeared in the 2020s, including the Russian invasion of Ukraine in 2022, the Hamas attack on Israel in 2023, and the difficulty of converting from vehicles with internal combustion engines to electric vehicles (Fanchi, 2024).

The development of nuclear fusion energy is a wild card in this discussion. The harnessing of nuclear fusion would provide a nearly limitless and sustainable source of energy. The commercialization of nuclear fusion could be achieved by the middle of the 21st century, but Charles Seife has discussed ongoing issues such as cost overruns and a timeline that keeps changing in the development of commercial nuclear fusion prototype ITER (Seife, 2023).

References

AI Biden, 2023. FACT SHEET: President Biden Issues Executive Order on Safe, Secure, and Trustworthy Artificial Intelligence. White House. Published October 30, 2023. Retrieved from https://www.whitehouse.gov/briefing-room/statements-releases/2023/10/30/fact-sheet-president-biden-issues-executive-order-on-safe-secure-and-trustworthy-artificial-intelligence/ (accessed November 5, 2023).

Alexander, P.E. and K. Heckenlively, 2023. *Presidential Takedown: How Anthony Fauci, the CDC, NIH, and the WHO Conspired to Overthrow President Trump*. Skyhorse Publishing (November 15, 2022), New York.

Allen, G. and L. Abraham, 1971. *None Dare Call It Conspiracy*. Buccaneer Books, Cutchogue, New York.

Allison, G., 2015. The Thucydides Trap: Are the U.S. and China Headed for War? *The Atlantic*. Published September 24, 2015.

Abnett, K., 2022. EU parliament backs labelling gas and nuclear investments as green. *Thomson Reuters*. Published July 6, 2022. Retrieved from https://www.reuters.com/business/sustainable-business/eu-parliament-vote-green-gas-nuclear-rules-2022-07-06/ (accessed October 16, 2023).

Amaro, S., 2021. Nord Stream 2: The Russian pipeline that everybody's talking about. CNBC. Published October 18, 2021. Retrieved from https://www.cnbc.com/2021/10/18/nord-stream-2-the-russian-pipeline-that-everybodys-talking-about.html (accessed October 16, 2023).

Bailey, R., 1997. Who is Maurice Strong? *The National Review*. Published September 1, 1997.

Bakewell, C.M., 1901 (July). A Democratic Philosopher and His Work. Thomas Davidson: Born October 25, 1840. Died September 14, 1900. *International Journal of Ethics*, Volume 11.

Ball, M., 2021. The Secret History of the Shadow Campaign That Saved the 2020 Election. *Time Magazine*. February 4 issue.

Beavers, O., J. Thomsen, and B. Samuels, 2019. Mueller Probe: A Timeline from Beginning to End. *The Hill*. Published March 24, 2019. Retrieved from https://thehill.com/policy/national-security/435547-mueller-probe-a-timeline-from-beginning-to-end/ (accessed October 30, 2023).

Beck, G. and J. Haskins, 2022. *The Great Reset: Joe Biden and the Rise of Twenty-First-Century Fascism*. Contributor D. Kendal. Forefront Books, Simon & Schuster, New York.

Beck, G. and J. Haskins, 2023. *Dark Future: Uncovering the Great Reset's Terrifying Next Phase*. Contributor D. Kendal. Forefront Books, Simon & Schuster, New York.

Biden, J., 2021. FACT SHEET: President Biden Sets 2030 Greenhouse Gas Pollution Reduction Target Aimed at Creating Good-Paying Union Jobs and Securing U.S. Leadership on Clean Energy Technologies. United States White House. Published April 22, 2021. Retrieved from https://www.whitehouse.gov/briefing-room/statements-releases/2021/04/22/fact-sheet-president-biden-sets-2030-greenhouse-gas-pollution-reduction-target-aimed-at-creating-good-paying-union-jobs-and-securing-u-s-leadership-on-clean-energy-technologies/ (accessed August 30, 2023).

Bill of Rights, 1789. The Bill of Rights: A Transcription. National Archives. Retrieved from https://www.archives.gov/founding-docs/bill-of-rights-transcript (accessed September 27, 2023).

Blinken, A.J., 2021 (February 19). The United States Officially Rejoins the Paris Agreement. United States Department of State. Retrieved from https://www.state.gov/the-united-states-officially-rejoins-the-paris-agreement/ (accessed August 26, 2023).

Bobb, C., 2023. *Stealing Your Vote: The Inside Story of the 2020 Election and What It Means for 2024*. Skyhorse Publishing, New York.

Bocking, S., 2012. Nature on the Home Front: British Ecologists' Advocacy for Science and Conservation. *Environment and History*, June issue, doi: 10.3197/096734012X13303670112894. Retrieved from https://www.researchgate.net/profile/Stephen_Bocking/publication/233510180_Nature_on_the_Home_Front_British_Ecologists%27_Advocacy_for_Science_and_Conservation/links/568ab87008aebccc4e1a1819/Nature-on-the-Home-Front-British-Ecologists-Advocacy-for-Science-and-Conservation.pdf (accessed July 31, 2017).

Boden, T.A., G. Marland, and R.J. Andres, 2017. National CO_2 Emissions from Fossil-Fuel Burning, Cement Manufacture, and Gas Flaring: 1751-2014. Data from U.S. EPA website https://www.epa.gov/ghgemissions/global-greenhouse-gas-overview (accessed December 11, 2022).

BOE 4IR, 2023. The Fourth Industrial Revolution. First written by Klaus Schwab in 2018. Editors of the Encyclopedia Britannica. Published May 31, 2023. Retrieved from https://www.britannica.com/topic/The-Fourth-Industrial-Revolution-2119734 (accessed September 25, 2023).

BOE AL Strong, 2023. Anna Louise Strong. Editors of the *Encyclopedia Britannica*. Retrieved from https://www.britannica.com/biography/Anna-Louise-Strong (accessed September 8, 2023).

BOE Ardern, 2023. Jacinda Ardern. *Britannica Online Encyclopedia*. Published October 27, 2023. Retrieved from https://www.britannica.com/biography/Jacinda-Ardern/The-2017-election (accessed November 2, 2023).

BOE Comey, 2023. James Comey. Editors of the Encyclopedia Britannica. Published May 1, 2023. Retrieved from https://www.britannica.com/biography/James-Comey (accessed October 30, 2023).

BOE Congress of Vienna, 2022. Congress of Vienna. Britannica Online Encyclopedia. Published December 11, 2022. Retrieved from https://www.britannica.com/event/Congress-of-Vienna (accessed May 18, 2023).

BOE Gaza Strip, 2023. Gaza Strip. Britannica Online Encyclopedia. Published October 28, 2023. Retrieved from https://www.britannica.com/place/Gaza-Strip (accessed October 28, 2023).

BOE Giuliani, 2023. Rudy Giuliani. Editors of the Encyclopedia Britannica. Published October 24, 2023. Retrieved from https://www.britannica.com/biography/Rudy-Giuliani (accessed October 30, 2023).

BOE Marshall Plan, 2023. Marshall Plan. Britannica Online Encyclopedia. Published March 29, 2023. Retrieved from https://www.britannica.com/event/Marshall-Plan (accessed November 2, 2023).

BOE Meiji Restoration, 2023. Meiji Restoration. Britannica Online Encyclopedia. Published March 29, 2023. Retrieved from https://www.britannica.com/event/Meiji-Restoration (accessed May 18, 2023).

BOE New Left, 2023. New Left Political Movement. Britannica Online Encyclopedia. Published June 5, 2023. Retrieved from https://www.britannica.com/topic/New-Left (accessed September 27, 2023).

BOE Opium Wars, 2023. Opium Wars. Britannica Online Encyclopedia. Published April 28, 2023. Retrieved from https://www.britannica.com/topic/Opium-Wars (accessed May 18, 2023).

BOE Russian Revolution, 2023. Russian Revolution. Britannica Online Encyclopedia. Published April 24, 2023. Retrieved from https://www.britannica.com/event/Russian-Revolution (accessed May 19, 2023).

BOE Standard Oil, 2023. Standard Oil Company. Editors of the Encyclopedia Britannica. Retrieved from https://www.britannica.com/topic/Standard-Oil (accessed September 19, 2023).

BOE Thirty Years War, 2023. Thirty Years' War. Britannica Online Encyclopedia. Published April, 1, 2023. Retrieved from https://www.britannica.com/event/Thirty-Years-War (accessed May 18, 2023).

BOE Versailles, 2023. Treaty of Versailles. Britannica Online Encyclopedia. Published April 24, 2023. Retrieved from https://www.britannica.com/event/Treaty-of-Versailles-1919 (accessed May 18, 2023).

BOE Webb, 2017. Sidney and Beatrice Webb. Britannica Online Encyclopedia. Published November 15, 2017. Retrieved from https://www.britannica.com/biography/Sidney-and-Beatrice-Webb (accessed July 26, 2023).

BOE Westphalia, 2023. Peace of Westphalia. Britannica Online Encyclopedia. Published May 12, 2023. Retrieved from https://www.britannica.com/event/Peace-of-Westphalia (accessed May 18, 2023).

BOE World War I, 2023. World War I. Britannica Online Encyclopedia. Published May 10, 2023. Retrieved from https://www.britannica.com/event/World-War-I (accessed May 18, 2023).

Boisvert, N., 2022. Trudeau Ends Use of Emergencies Act, Says "Situation Is No Longer an Emergency". *CBC News*. Posted online February 23, 2022. Retrieved from https://www.cbc.ca/news/politics/trudeau-event-feb23-1.6361847 (accessed November 3, 2023).

Boyd, J., 2020. Joe Biden Just Promised No Coal Power Plants Will Ever Be Built in America If He Wins. The Federalist, Published September 30, 2020. Retrieved from https://thefederalist.com/2020/09/30/joe-biden-just-promised-no-coal-power-plants-will-ever-be-built-in-america-if-he-wins/ (accessed September 15, 2023).

Bur, J., 2021. Answers to Federal Workers' Questions about Government Labor unions. Federaltimes.com. Retrieved from https://www.federaltimes.com/management/hr/2021/09/01/answers-to-federal-workers-questions-about-government-labor-unions/ (accessed October 23, 2023).

Brzezinski, Z., 2004. Special Address to the 2004 Trilateral Commission Plenary Meeting in Warsaw.

Burnham, J., 1941. The Managerial Revolution. First published by John Day Company, New York. Reprinted in 1972 by Greenwood Press, Westport, CT.

CARB, 2023. Advanced Clean Cars II Regulations: All New Passenger Vehicles Sold in California to be Zero Emissions by 2035. California Air Resources Board. Retrieved from https://ww2.arb.ca.gov/our-work/programs/advanced-clean-cars-program/advanced-clean-cars-ii (accessed October 2, 2023).

Carlson, T., 2018. *Ship of Fools: How a Selfish Ruling Class Is Bringing America to the Brink of Revolution*. Free Press, Simon & Schuster, New York.

Carson, R. 1962. *Silent Spring*. Houghton Mifflin, New York.

CDC COVID-19, 2022. CDC Museum COVID-19 Timeline. United States Center for Disease Control and Prevention. Retrieved from (updated

March 15, 2023) https://www.cdc.gov/museum/timeline/covid19.html (accessed October 9, 2023).
CFR BRI, 2021. China's Belt and Road – Implications for the United States. Council on Foreign Relations, Independent Task Force Report No. 79 (March). J.J. Lew and G. Roughead, Chairs; J. Hillman and D. Sacks, Project Directors; Richard Haas, CFR President. Retrieved from https://www.cfr.org/report/chinas-belt-and-road-implications-for-the-united-states/ (accessed October 17, 2023).
CFR BRICS, 2017. How the BRICS Got Here. Expert Brief by Alyssa Ayres Published August 31, 2017. Council on Foreign Relations. Retrieved from https://www.cfr.org/expert-brief/how-brics-got-here (accessed October 18, 2023).
Chaffetz, J., 2018. *The Deep State*. Broadside Books, HarperCollins, New York.
Chaffetz, J., 2023. *The Puppeteers: The People Who Control the People Who Control America*. Broadside Books, HarperCollins, New York.
Chatham House, 2023. Our History: 1919. Chatham House. Retrieved from https://www.chathamhouse.org/about-us/our-history (accessed October 7, 2023).
Chen, C., D. Li, Y. Li, S. Piao, X. Wang, M. Huang, P. Gentine, R. R. Nemani, and R.B. Myneni, 2020. Biophysical Impacts of Earth Greening Largely Controlled by Aerodynamic Resistance. *Science Advances*, 6, 9.
Clark, K., 1964. *John Ruskin — Selected Writings*. Penguin Books, London printed in 1991. First Published as Ruskin Today in 1964 by Holt, Rinehart and Winston, New York.
CLINTEL, 2023. Climate Intelligence. Retrieved from https://clintel.org/ (accessed September 2, 2023).
CLINTEL Declaration, 2023. World Climate Declaration. PDF dated August 14, 2023. Retrieved from https://clintel.org/wp-content/uploads/2023/08/WCD-version-081423.pdf (accessed September 2, 2023).
Colbeck, P., 2022. *The 2020 Coup: What Happened. What We Can Do*. McHenry Press at https://mchenrypress.com/ (accessed December 1, 2023).
Collins, L., 2009. The Truth about Tytler. Posted online January 25, 2009. Retrieved from http://www.lorencollins.net/tytler.html (accessed June 12, 2023).
Cook, J., D. Nuccitelli, S.A. Green, M. Richardson, B. Winkler, R. Painting, R. Way, P. Jacobs, and A. Skuce, 2013. Quantifying the consensus on anthropogenic global warming in the scientific literature. *Environmental Research Letters*, Volume 8, pp. 1–7. doi:10.1088/1748-9326/8/2/024024.
Cruz, T., 2023. *Unwoke: How to Defeat Cultural Marxism in America*. Regnery Publishing, Washington, DC.

Daily Mail Online, 2021. Stern Words, Economic Sanctions and a Diplomatic Boycott: The Action Obama Took to Deter Putin from Invading Crimea in 2014. *Daily Mail Online*. Retrieved from (updated December 7, 2021) https://www.dailymail.co.uk/news/fb-10285787/The-action-Obama-took-deter-Putin-invading-Crimea-2014.html (accessed October 16, 2023).

Dalio, R., 2021. *The Changing World Order*. Simon & Schuster, New York.

Duffy, M., 1974. Who's Who: Walter Lippmann. Retrieved from http://www.firstworldwar.com/bio/lippmann.htm (accessed March 14, 2018).

Earth Charter, 2023. The Earth Charter. Retrieved from http://earthcharter.org/discover/download-the-charter/ (accessed September 10, 2023).

ECI–Strong, 2015. Tribute to Maurice F. Strong (1929–2015). Earth Charter Initiative. Earth Charter International Secretariat and Council. Posted online November 30, 2015. Retrieved from https://earthcharter.org/tribute-to-maurice-f-strong-1929-2015/ (accessed September 10, 2023).

Eisenhower, D.D., 1961. President Dwight D. Eisenhower's Farewell Address (1961). National Archives. Farewell Address to the Nation. Published January 17, 1961. Retrieved from https://www.archives.gov/milestone-documents/president-dwight-d-eisenhowers-farewell-address (accessed September 24, 2023).

Elton, C., 1958. *The Ecology of Invasions by Animals and Plants*. Reprinted by University of Chicago Press, Chicago, IL, in 2000.

Energy Institute Stats, 2023. Primary Energy: Consumption by fuel. Statistical Review of World Energy. Retrieved from https://www.energyinst.org/statistical-review (accessed October 9, 2023).

Engels, F., 1886. Ludwig Feuerbach and the End of Classical German Philosophy. *Die Neue Zeit*. Retrieved from https://www.marxists.org/archive/marx/works/download/Marx_Ludwig_Feurbach_and_the_End_of_German_Classical_Philosop.pdf (accessed July 26, 2017).

ENTSO-E, 2022. A Power System for a Carbon Neutral Europe. European Network of Transmission System Operators for Electricity. Published October 10, 2022. Retrieved from https://vision.entsoe.eu/ (accessed October 14, 2023).

Epstein, A.J., 2014. *The Moral Case for Fossil Fuels*. Penguin Random House, New York.

Epstein, A.J., 2022. *Fossil Future*. Penguin Random House, New York.

EU 2050, 2023. 2050 Long-Term Strategy. European Commission Climate Action. Retrieved from https://ec.europa.eu/clima/eu-action/climate-strategies-targets/2050-long-term-strategy_en (accessed October 14, 2023).

Fabian Society, 2023. About the Fabian Society. Fabian Society website https://fabians.org.uk/about-us/ (accessed July 22, 2023).

Fanchi, J.R. and Fanchi, C.J., 2015. *The Role of Oil and Natural Gas in the Global Energy Mix*. Volume 3: Oil and Natural Gas, Energy Science and

Technology, Executive Editor: Dr. J N Govil, ISBN: 1-62699-072-7, Studium Press, New Delhi, India.

Fanchi, J.R., 2019. *The Goldilocks Policy: The Basis for a Grand Energy Bargain.* World Scientific, Singapore.

Fanchi, J.R., 2023. *Confronting the Enigma of Time.* World Scientific, Singapore.

Fanchi, J.R., 2024. *Energy in the 21st Century,* 5th Edition. World Scientific, Singapore.

FBI Comey Remarks, 2016. Statement by FBI Director James B. Comey on the Investigation of Secretary Hillary Clinton's Use of a Personal E-Mail System. Federal Bureau of Investigation website https://www.fbi.gov/news/press-releases/statement-by-fbi-director-james-b-comey-on-the-investigation-of-secretary-hillary-clinton2019s-use-of-a-personal-e-mail-system (accessed October 30, 2023).

Flint, J., 1974. *Cecil Rhodes.* Little, Brown and Company, Boston.

FLRA EO10988, 2023. 50th Anniversary: Executive Order 10988. Federal Labor Relations Authority. FLRA.gov. Retrieved from https://www.flra.gov/50th_Anniversary_EO10988 (accessed October 23, 2023).

Fontinelle, A., C. Clarke, and D. Costagliola, 2022. Paris Agreement/COP21. Investopedia.com. Retrieved from https://www.investopedia.com/terms/p/paris-agreementcop21.asp (accessed August 26, 2023).

Fosdick, R.B., 1952. *The story of the Rockefeller Foundation.* Harper, New York.

Founders, 1776. Declaration of Independence. U.S. Department of the Interior. Retrieved from http://www.constitution.org/us_doi.pdf (accessed July 28, 2017).

Franklin, B., 1755. Pennsylvania Assembly: Reply to the Governor, November 11, 1755. National Archives. Retrieved from https://founders.archives.gov/documents/Franklin/01-06-02-0107 (accessed November 11, 2023).

Frazin, R., 2022. Biden Blowback Shows Coal Still Has Sway in US Politics. *The Hill.* Published September 7. Retrieved from https://thehill.com/policy/energy-environment/3723927-biden-blowback-shows-coal-still-has-sway-in-us-politics/ (accessed September 16, 2023).

Garrow, David J., 1994. *Liberty and Sexuality: The Right to Privacy and the Making of Roe v. Wade.* Macmillan, New York.

Gavrilis, G., 2021. *The Council on Foreign Relations: A Short History.* The Council on Foreign Relations, New York.

Glasov, J., editor, 2023. *Obama's True Legacy: How He Transformed America.* Republic Book Publishers, New York.

Glubb, J.B., 1976. *The Fate of Empires and Search for Survival.* William Blackwood & Sons Ltd, Edinburgh, Scotland. Retrieved from https://archive.org/details/pdfy-2F_iHS6BLtGJb2ad (accessed June 12, 2023).

Goh, B. and R. Woo, 2022. COVID-HIT SHANGHAI TO END TWO-MONTH LOCKDOWN on June 1. *Thomson Reuters.* Published May 30, 2022.

Retrieved from https://www.reuters.com/world/china/some-beijing-back-work-shanghai-inches-closer-ending-covid-lockdown-2022-05-30/ (accessed November 2, 2023).

Golden, C.D., 2020. Watch: Obama Talks About Using a "Stand-in, a Front-Man" for a Third Term. *Western Journal*. Published December 15, 2020. Retrieved from https://www.westernjournal.com/watch-obama-talks-using-stand-front-man-third-term/ (accessed September 16, 2023).

Gonzalez, M. and K.C. Gorka, 2022. How Cultural Marxism Threatens the United States — and How Americans Can Fight It. The Heritage Foundation Special Report Number 262. Published November 14, 2022. Retrieved from https://www.heritage.org/progressivism/report/how-cultural-marxism-threatens-the-united-states-and-how-americans-can-fight (accessed December 7, 2023).

Gore, A., 2009. *Our Choice: A Plan to Solve the Climate Crisis*. Rodale-Melchir Media, New York.

Gore-Nobel Prize, 2007. The Nobel Peace Prize 2007. Retrieved from https://www.nobelprize.org/prizes/peace/2007/summary/ (accessed September 13, 2023).

Gore-Oscars, 2007. The 79th Academy Awards. Retrieved from https://www.oscars.org/oscars/ceremonies/2007 (accessed September 13, 2023).

Haass, R., 2020. *The World: A Brief Introduction*. Penguin Press, New York.

Hanson, V.D., 2021. *The Dying Citizen: How Progressive Elites, Tribalism, and Globalization Are Destroying the Idea of America*. Basic Books, New York.

Hayek, F., 1944. *The Road to Serfdom*. Routledge, London.

Hemingway, M. 2021a. "Zuckerbucks" and the 2020 Election. Imprimis. Published October 2021. Retrieved from https://imprimis.hillsdale.edu/zuckerbucks-2020-election/ (accessed October 31, 2023).

Hemingway, M. 2021b. *Rigged: How the Media, Big Tech, and the Democrats Seized Our Elections*. Regnery Publishing, Standard Edition, Washington, DC. Published October 12, 2021.

History.com New Republic, 2020. This Day in History: November 7, 1914. First Issue of *The New Republic*. Published in History.com. Retrieved from https://www.history.com/this-day-in-history/first-issue-of-the-new-republic-Published (accessed October 30, 2023).

History.com Trump Impeached, 2021. This Day in history: December 18, 2019. President Donald Trump Impeached. History.com. Retrieved from https://www.history.com/this-day-in-history/president-trump-impeached-house-of-representatives (accessed October 30, 2023).

Horowitz, D., 2018. *Dark Agenda: The War to Destroy Christian America*. Humanix Books, West Palm Beach, FL.

Horowitz, D., 2022. *Final Battle: The Next Election Could Be the Last*. Humanix Books, West Palm Beach, FL.

Hourdin, F., T. Mauritsen, A. Gettelman, J.C. Golaz, V. Balaji, Q. Duan, D. Folini, D., Ji, D. Klocke, Y. Qian, F. Rauser, C. Rio, L. Tomassini, M. Watanabe, and D. Williamson, 2017. The Art and Science of Climate Model Tuning. *Bulletin of the American Meteorological Society*, March, 589–602.

Hu, Y., Y. Tang, K. Wang, and X. Yang, 2022. Low Carbon and Economic Dispatching of Electric-Gas Integrated Energy System with Liquid Storage Carbon Capture Equipment. *Frontiers in Energy Research*, 11 pages. doi:10.3389/fenrg.2022.986646.

Huntington, S.P., 1996. *The Clash of Civilizations*. Simon and Schuster, London.

Igini, M., 2023. Why Electric Cars Are Better for the Environment. earth.org. Retrieved from https://earth.org/electric-cars-environment/ as of February 16, 2023 (accessed October 14, 2023).

Investopedia BRICS, 2023. BRICS: Acronym for Brazil, Russia, India, China, and South Africa by J. Chen, September 29, 2023. Investopedia. Retrieved from https://www.investopedia.com/terms/b/brics.asp (accessed October 19, 2023).

ITER, 2023. Fusion Energy. *International Thermonuclear Experimental Reactor*. Retrieved from http://www.iter.org (accessed June 10, 2023).

Jefferson, T., 1802. Jefferson's Wall of Separation Letter. U.S. Constitution Retrieved from https://usconstitution.net/jeffwall.html (accessed September 27, 2023).

Johnson, M., 1958. Color, Communism, and Common Sense. Retrieved from https://archive.org/details/color-communism-and-common-sense (accessed September 29, 2023).

Jones, A., 2022. *The Great Reset*. Skyhorse Publishing, New York.

Jones, A., and K. Heckenlively, 2023. *The Great Awakening: Defeating the Globalists and Launching the Next Great Renaissance*. Skyhorse Publishing, New York.

Katella, K., 2021. Our Pandemic Year – A COVID-19 Timeline. *Yale Medicine*. Retrieved from https://www.yalemedicine.org/news/covid-timeline (accessed October 25, 2023).

Klare, M.T., 2001. *Resource Wars: The New Landscape of Global Conflict*. Henry Holt and Company, New York.

Klare, M.T., 2004. *Blood and Oil*. Henry Holt and Company, New York.

Koonin, S. 2021. *Unsettled: What Climate Science Tells Us, What It Doesn't, and Why It Matters*. BenBella Books, Dallas, TX.

Kotkin, J., 2019. America's Drift toward Feudalism. *American Affairs*, III, Number 4 (Winter), 96–107.

Kotkin, J., 2020. Neo-Feudalism in California. *American Affairs*, IV(2) (Summer), 62–77.

Kragh, H., 2004. *Matter and Spirit in the Universe*. Imperial College Press, London.

Kurlantzick, J., 2022. China's Collapsing Global Image — How Beijing's Unpopularity Is Undermining Its Strategic, Economic, and Diplomatic Goals. Council on Foreign Relations Discussion Paper (July). Retrieved from https://cdn.cfr.org/sites/default/files/report_pdf/Kurlantzick_DP_Chinas CollapsingGlobalImage.pdf (accessed October 17, 2023).

Lankester, E.R., 1913. *The Effacement of Nature by Man*. Chapter XXVIII in Science from an Easy Chair; A Second Series. H. Holt and Company, New York.

Leake, J., and P.A. McCullough, 2023. *The Courage to Face COVID-19: Preventing Hospitalization and Death While Battling the Bio-Pharmaceutical Complex*. Skyhorse (November 22, 2022), New York.

Le Grand, C., 2022. "Fractured" Pandemic Response Failed the Most Vulnerable, Independent Report Finds. *The Sydney Morning Herald*. Published October 20, 2022. Retrieved from https://www.smh.com.au/national/fractured-pandemic-response-failed-the-most-vulnerable-independent-report-finds-2022 1018-p5bqso.html (accessed November 3, 2023).

Lenin, V.I., 1901. The Abolition of the Antithesis between Town and Country. Particular Questions Raised by the "Critics". First published as Chapter IV in *The Agrarian Question and the "Critics of Marx"* by Zarya, December 1901; Lenin Collected Works, Foreign Languages Publishing House, 1961, Moscow, Volume 5, pp. 103–222. Retrieved from https://www.marxists.org/archive/lenin/works/1901/agrarian/index.htm (accessed September 6, 2017).

Lenin CEC, 1918. The Fundamental Law of Land Socialization. Decree of the Central Executive Committee, February 19, 1918. Website https://www.marxists.org/history/ussr/events/revolution/documents/1918/02/19.htm (accessed September 6, 2017).

Levin, M., 2021. *American Marxism*. Threshold Editions, Simon & Schuster, New York.

Levin, M., 2023. *The Democrat Party Hates America*. Threshold Editions, Simon & Schuster, New York.

Lewis, J., 1985. The Birth of the EPA. *EPA Journal*. Retrieved from https://archive.epa.gov/epa/aboutepa/birth-epa.html (accessed September 18, 2017).

Lindsey, R. and L. Dahlman, 2022. Climate Change: Global Temperature. United States National Oceanic and Atmospheric Administration. Retrieved from (last updated January 18, 2023) https://www.climate.gov/news-features/understanding-climate/climate-change-global-temperature (accessed October 9, 2023).

Locke, J., 1680. *Two Treatises of Government*. The Easton Press, Norwalk, CT. Books that Changed the World, 1991.

Locke, J., 1690. An Essay Concerning Humane Understanding. The Project Gutenberg ebook. Website https://www.gutenberg.org/files/10615/10615-h/10615-h.htm (accessed July 27, 2017).

Lofgren, M., 2014. Essay: Anatomy of the Deep State. Retrieved from http://billmoyers.com/2014/02/21/anatomy-of-the-deep-state/ (accessed October 2, 2023).

Lomborg, B., 2001. *The Skeptical Environmentalist*. Cambridge University Press, Cambridge, United Kingdom.

Lomborg, B., 2021. *False Alarm*, Updated Edition. Basic Books, New York.

Lovelock, J., 2009. *The Vanishing Face of GAIA*. Basic Books, New York.

MacNeill, J., P. Winsemius, and T. Yakushiji, 1991. *Beyond Interdependence*. A Trilateral Commission Book. Oxford University Press, New York.

Madison, J., 1792. Property. *The Founders' Constitution*, edited by P.B. Kurland and R. Lerner. Published by University of Chicago Press, Chicago, IL, Volume 1, 1987, pp. 598–599.

Malone, R.W., 2023. *Lies My Gov't Told Me: And the Better Future Coming (Children's Health Defense)*. Skyhorse Publishing (December 6, 2022), New York.

Manitou-Rio, 2020. Maurice Strong's Opening Statement to the Rio Summit (June 3, 1992). Manitou Foundation. Retrieved from https://mauricestrong.net/index.php?option=com_content&view=article&id=36:rio2&catid=13&Itemid=59 (accessed September 9, 2023).

Manitou–Stockholm, 2020. United Nations Conference on the Human Environment, 1972. Opening Statement by Maurice Strong, Secretary-General of the Conference. Manitou Foundation. Retrieved from https://mauricestrong.net/index.php?option=com_content&view=article&id=103:stockholm&catid=13&Itemid=59 (accessed September 8, 2023).

Manitou–Strong, 2020. Maurice Strong: Short Biography. Manitou Foundation. Retrieved from https://www.mauricestrong.net/index.php?option=com_content&view=article&id=15&Itemid=24 (accessed September 8, 2023).

Märald, E., 2002. Everything Circulates: Agricultural Chemistry and Recycling Theories in the Second Half of the Nineteenth Century. *Environment and History*, Volume 8, pp. 65–84.

Marks, J., 2023. Gaza: The History That Fuels the Conflict. History. Retrieved from https://www.history.com/news/gaza-conflict-history-israel-palestine (accessed October 28, 2023).

Marx, K., 1841. The Difference between the Democritean and Epicurean Philosophy of Nature. First published in 1902; obtained from *Marx-Engels Collected Works*, Volume 1 by Progress Publishers. Website https://ia801307.us.archive.org/19/items/Marx_Karl_-_Doctoral_Thesis_-_The_Difference_Between_the_Democritean_and_Epicure/Marx_Karl_-_Doctoral_Thesis_-_The_Difference_Between_the_Democritean_and_Epicurean_Philosophy_of_Nature.pdf (accessed July 26, 2017).

Marx, K., 1845. Theses on Feurbach. Retrieved from http://www.marx2mao.com/M&E/TF45.html (accessed September 8, 2017).

Marx, K., 1866. February 13 Letter from Marx to Engels. Marx & Engels Collected Works, Volume 42, Letters 1864-68, Copyright 2010 by Lawrence & Wishart, Electric Book, ISBN 978-1-84327-986-0. Retrieved from http://www.hekmatist.com/Marx%20Engles/Marx%20&%20Engels%20Collected%20Works%20Volume%2042_%20Ka%20-%20Karl%20Marx.pdf (accessed July 29, 2017).

Marx, K., 1887. *Capital: A Critique of Political Economy*, Volume I. First English edition with Fourth German edition changes. Retrieved from https://www.marxists.org/archive/marx/works/download/pdf/Capital-Volume-I.pdf (accessed July 29, 2017).

Marx, K., 1894. *Capital: A Critique of Political Economy*, Volume III. Edited by Friedrich Engels and completed by Engels 11 years after Marx's death. Retrieved from https://www.marxists.org/archive/marx/works/download/pdf/Capital-Volume-III.pdf (accessed July 29, 2017).

Marx, K., and Engels, F., 1848. Manifesto of the Communist Party. *Communist Manifesto and Other Writings*, translated by Martin Milligan, The Easton Press Collector's Edition, Norwalk, CT, 2005.

Mayflower Compact, 1620. The Mayflower Compact. The Mayflower Society. Retrieved from https://themayflowersociety.org/history/the-mayflower-compact/ (accessed November 11, 2023).

Mayo Clinic COVID-19, 2023. 2020 — COVID-19 and Related Vaccine Development and Research. The Mayo Clinic. Retrieved from https://www.mayoclinic.org/diseases-conditions/history-disease-outbreaks-vaccine-timeline/covid-19 (accessed October 25, 2023).

Meadows, D.H., D.L. Meadows, J. Randers, and W.W. Behrens III,1972. *Limits to Growth*. Potomac Associates, Washington, DC.

Mische, P.M. and M.A. Ribeiro. 1998. Ecological Security and the United Nations System. Chapter 10 in *The Future of the United Nations: Potential for the Twenty-first Century*, edited by C.F. Alger, United Nations University Press, New York.

Moreno, P., 2011. The History of Public-Sector Unionism. Hillsdale College. Retrieved from https://www.hillsdale.edu/educational-outreach/free-market-forum/2011-archive/the-history-of-public-sector-unionism/ (accessed October 23, 2023).

Mosher, S.W., 2017. *Bully of Asia — Why China's Dream is the New Threat to World Order*. Regnery Publishing, Washington, DC.

Murray, D., 2022. *The War on the West*. HarperCollins Publishers Australia Pty. Ltd., Sydney, Australia.

Murray, R.L., 2001. *Nuclear Energy: An Introduction to the Concepts, Systems, and Applications of Nuclear Processes*, 5th Edition. Butterworth-Heinemann, Boston, MA.

NASA Climate Change, 2023. Scientific Consensus: Earth's Climate is Warming. Most recent update August 24, 2023. Retrieved from https://climate.nasa.gov/scientific-consensus/ (accessed September 2, 2023).

NIPCC CCR-II, 2013. Climate Change Reconsidered II – Physical Science Summary for Policy Makers. Report of the Nongovernmental International Panel on Climate Change. Retrieved from http://climatechangereconsidered.org/about-the-nipcc/ (accessed August 29, 2023).

NOAA Sea Level, 2023 (January 20). Is sea level rising? National Ocean Service, National Oceanographic and Atmospheric Administration. Retrieved from https://oceanservice.noaa.gov/facts/sealevel.html (accessed August 26, 2023).

NOAA Storm Surge, 2023. Storm Surge Overview. National Oceanic and Atmospheric Administration National Hurricane Center and Central Pacific Hurricane Center. Retrieved from http://www.nhc.noaa.gov/surge/ (accessed August 26, 2023).

Noble, S.M., M. Mende, D. Grewal, and A. Parasuraman, 2022. The Fifth Industrial Revolution: How Harmonious Human-Machine Collaboration is Triggering a Retail and Service [R]evolution. *Journal of Retailing*, 98, 199–208.

Nolan, P. 2021. What Are Environmental, Social, and Governance (ESG) Criteria? Last updated November 11, 2021. Retrieved from https://www.thebalancemoney.com/what-are-environmental-social-and-governance-esg-criteria-5112974 (accessed October 19, 2023).

NRDC, 2022. Electric Vehicle Charging Explained. National Resource Defense Council. Published July 5, 2022. Retrieved from https://www.nrdc.org/stories/electric-vehicle-charging-explained (accessed October 14, 2023).

NWE-Rhodes, 2018. Cecil Rhodes. *New World Encyclopedia*. Retrieved from http://www.newworldencyclopedia.org/entry/Cecil_Rhodes#cite_note-9 (accessed February 20, 2018).

OGCI Founding, 2023. Oil and Gas Climate Initiative. Retrieved from http://ogci.com/ (accessed August 30, 2023).

OGCI Strategy, 2023. OGCI's Strategy and Principles. Retrieved from https://www.ogci.com/about/strategy-and-principles (accessed August 30, 2023).

Ordoñez. F., 2020. What Most Biden Picks Have In Common: Time In Obama Administration. Biden Transition Updates Published December 12, 2020. National Public Radio. Retrieved from https://www.npr.org/sections/biden-transition-updates/2020/12/12/945627799/what-most-biden-picks-have-in-common-time-in-obama-administration (accessed September 16, 2023).

Orwell, G., 1937. *The Road to Wigan Pier*. Victor Gollancz Ltd., London, United Kingdom.

Orwell, G., 1945. *Animal Farm*. First Published 17 August 1945. Secker and Warburg, London, United Kingdom.

Orwell, G., 1946. Second Thoughts on James Burnham. *Polemic Magazine* 3, May 1946.

Orwell, G., 1949. *1984 – A Novel*. Secker & Warburg, London, United Kingdom.

OWHA Climate, 2023. President Obama on Climate and Energy. Retrieved from https://obamawhitehouse.archives.gov/sites/obamawhitehouse.archives.gov/files/achievements/theRecord_climate_0.pdf (accessed September 14, 2023).

Parks, M., 2022. Congress May Change This Arcane Law to Avoid Another, January 6. NPR. Published January 8, 2022. Retrieved from https://www.npr.org/2022/01/08/1071239044/congress-may-change-this-arcane-law-to-avoid-another-jan-6 (accessed October 30, 2023).

Parks, M., Dec. 2022. Congress Passes Election Reform Designed to Ward Off Another, January 6. NPR. Published December 23, 2022. Retrieved from https://www.npr.org/2022/12/22/1139951463/electoral-count-act-reform-passes (accessed October 30, 2023).

Patel, K., 2023. *Government Gangsters: The Deep State, the Truth, and the Battle for Our Democracy*. Post Hill Press, New York.

Paul, R., 2023. *Deception: The Great Covid Cover-Up*. Regnery Publishing (October 10, 2023), Washington, DC.

Pease, E.R., 1916. *The History of the Fabian Society*. E.P. Dutton & Company, New York.

Peou, S., 2022. *Global Public Governance: Toward World Government?* World Scientific, Singapore.

Pifer, S., 2014. The Budapest Memorandum and U.S. Obligations. Brookings Institution article Published December 4, 2014. Retrieved from https://www.brookings.edu/blog/up-front/2014/12/04/the-budapest-memorandum-and-u-s-obligations/ (accessed June 9, 2023).

Plucinska, J., 2022. Nord Stream Gas "Sabotage": Who's Being Blamed and Why? Thomson Reuters. Published October 6, 2022. Retrieved from https://www.reuters.com/world/europe/qa-nord-stream-gas-sabotage-whos-being-blamed-why-2022-09-30/ (accessed October 16, 2023).

Prentis, H.W. Jr., 1943. The Cult of Competency. The text appeared in The General Magazine and Historical Chronicle (Volume XLV—Number III, April 1943) of the University of Pennsylvania; the publication was later merged with *The Pennsylvania Gazette*, the alumni magazine of the University of Pennsylvania. Retrieved from http://ergo-sum.net/literature/CultOfCompetency.pdf (accessed June 13, 2023).

Quigley, C., 1961. *The Evolution of Civilizations: An Introduction to Historical Analysis*, First Edition. Macmillan, New York.

Quigley, C., 1966. *Tragedy and Hope – A History of the World in Our Time*. Macmillan, New York.

Quigley, C., 1981. *The Anglo-American Establishment*. Books in Focus, New York.

Reagan, R., 1961. Ronald Reagan address to the annual meeting of the Phoenix Chamber of Commerce, March 30, 1961.

Reuters BRICS, 2023. What Is BRICS and Who Are Its Members? by Bhargav Acharya. Thomson Reuters. Published August 21, 2023. Retrieved from https://www.usnews.com/news/world/articles/2023-08-21/factbox-what-is-brics-and-who-are-its-members (accessed October 19, 2023).

Reuters Canadian Truckers, 2022. Canada's Trudeau Invokes Emergency Powers in Bid to End Protests. The Standard. Published February 15, 2022. Retrieved from https://www.thestandard.com.hk/breaking-news/section/6/187145/Canada (accessed November 3, 2023).

Reuters EV Sales, 2023. US EV Market Struggles with Price Cuts and Rising Inventories by J. White and B. Klayman. Thomson Reuters. Published July 11, 2023. Retrieved from https://www.reuters.com/business/autos-transportation/slow-selling-evs-are-auto-industrys-new-headache-2023-07-11/ (accessed October 2, 2023).

Reuters Emergency Powers, 2022. Canada's Parliament Approves Trudeau's Emergency Powers. CGTN Published February 22, 2022. Retrieved from https://news.cgtn.com/news/2022-02-22/Canada-s-parliament-approves-Trudeau-s-emergency-powers-17QXNtywxtS/index.html (accessed November 3, 2023).

Reuters Wiretapping, 2017. Trump Campaign Adviser Was Wiretapped under Secret Court Orders: CNN. Thomson Reuters. Published September 18, 2017. Retrieved from https://www.reuters.com/article/us-usa-trump-russia-probe-idUSKCN1BU019 (accessed October 30, 2023).

RF–Mission, 2023. About Us: Mission and Vision. Rockefeller Foundation. Retrieved from https://www.rockefellerfoundation.org/about-us/mission-and-vision/ (accessed September 20, 2023).

Rizzi, B., 1939. *The Bureaucratization of the World.* 1939 French edition translated by A. Buick from French to English. Retrieved from https://www.marxists.org/archive/rizzi/bureaucratisation/index.htm (accessed October 2, 2023).

Roberts, S., 2015. Maurice Strong, Environmental Champion, Dies at 86. Published December 1, 2015 online at https://www.nytimes.com/2015/12/02/world/americas/maurice-strong-environmental-champion-dies-at-86.html, (accessed September 7, 2023).

Rockefeller, D., 1991. Foreword in Beyond Interdependence by MacNeill *et al.*

Rockefeller, D., 2002. *Memoirs.* Random House, New York.

Rockefeller, J.D., III, 1973. *The Second American Revolution.* HarperCollins, New York.

Rufo, C.F., 2023. *America's Cultural Revolution: How the Radical Left Conquered Everything.* Broadside Books, HarperCollins, New York.

RSV Bible, 1971. *Genesis* [Revised Standard Version Bible]. American Bible Society, New York.

Samuelson, E., 2017. Council on Foreign Relations: It's Secret British Origin Explained. The Millenium Report. Published November 29, 2017. Retrieved from https://themillenniumreport.com/2017/11/council-on-foreign-relations-its-secret-british-origin-explained/ (accessed July 24, 2023).

Santayana, G., 1905–1906. The Life of Reason or, The Phases of Human Progress. Volume I, *Reason in Common Sense*, Chapter XII: Flux and Constancy in Human Nature. Scribner's, New York, p. 284; see also https://en.wikisource.org/wiki/Reason_in_Common_Sense.

Sasse, G., 2017. Revisiting the 2014 Annexation of Crimea. Carnegie Endowment for International Peace, Carnegie Europe. Retrieved from https://carnegieeurope.eu/2017/03/15/revisiting-2014-annexation-of-crimea-pub-68423 (accessed October 16, 2023).

Schwab, K., 2020. Now Is the Time For A "Great Reset". World Economic Forum. Published June 3, 2020. Retrieved from https://www.weforum.org/agenda/2020/06/now-is-the-time-for-a-great-reset/ (accessed November 8, 2023).

Schweizer, P., 2020. *Profiles in Corruption: Abuse of Power by America's Progressive Elite*. HarperCollins, New York.

Schweizer, P., 2022. *Red-Handed: How American Elites Get Rich Helping China Win*. HarperCollins, New York.

Scripps Keeling, 2022. The Keeling Curve. Scripps Institution of Oceanography, University of California–San Diego. Retrieved from https://keelingcurve.ucsd.edu (accessed December 10, 2022).

Seife, C., 2023. World's Largest Fusion Project Is in Big Trouble, New Documents Reveal. Scientific American. Published June 15, 2023.

Shafer, P.W. and J.H. Snow, 1962. *The Turning of the Tides*. Long House Publishing, New Canaan, CT.

Shergold, P., J. Broadbent. I. Marshall, and P. Varghese, 2022. Fault Lines. Paul Ramsay Foundation. Retrieved from https://www.paulramsayfoundation.org.au/news-resources/fault-lines-an-independent-review-into-australias-response-to-covid-19 (accessed November 3, 2023).

Simkin, J., 2017. James Keir Hardie. Spartacus Educational article. Retrieved from http://spartacus-educational.com/PRhardie.htm (accessed March 11, 2018).

Skousen, W.C., 1970. *The Naked Capitalist*. Self-Published in Salt Lake City, UT.

Smil, V., 2003. *Energy at the Cross*roads. The MIT Press, Cambridge, MA.

Smith, H., 1991. *The World's Religions: Our Great Wisdom Traditions*. HarperCollins, New York. Published by Easton Press with the permission of HarperCollins in 2008.

Srivastava, D.K., and V.S. Ramamurthy, 2021. *Climate Change and Energy Options for a Sustainable Future*. World Scientific, Singapore.

Strong, M., 1998. A People's Earth Charter. Chairman of the Earth Council and Co-Chair of the Earth Charter Commission. Published March 5, 1998. Retrieved from https://earthcharter.org/wp-content/assets/virtual-library2/images/uploads/Maurice%20Strong%20on%20A-%20Peoples%20Earth%20Charter.pdf (accessed September 10, 2023).

Strong, M., 2000. *Where on Earth Are We Going?* Alfred A. Knopf, Toronto.

Sussman, B., 2012. *Eco-Tyranny*. WND Books, Washington, DC.

Tarbell, I.M., 1904. *The History of the Standard Oil Company*. McClure, Phillips and Company, New York.

de Tocqueville, A., 1966. *Democracy in America*. Volumes 1 and 2. HarperCollins Publishers, New York. Easton Press 2004 edition translated by George Lawrence. Norwalk, CT.

Torah, 1992. Genesis, *Torah*. The Easton Press Collector's Edition, Norwalk, CT.

Trotsky, L., 1939. The USSR in War. First published in *The New International (New York)*, Volume 5, Number 11, pp. 325–332. Leon Trotsky Internet Archive 2005. British edition published in Trotsky (1942).

Trotsky, L., 1942. *In Defense of Marxism*. First published in 1942 by Pioneer Publishers, online version transcribed by David Walters in 1998. Retrieved from https://www.marxists.org/archive/trotsky/idom/dm/dom.pdf (accessed October 2, 2023).

Troy, L., 1994. *The New Unionism*: Public Sector Unions in the Redistributive State. George Mason University Press, Fairfax.

U Thant, 1970. Human Environment and World Order. *International Journal of Environmental Studies*, 1(1–4), 13–17, doi: 10.1080/00207237008709390.

UN Charter, 1945. Charter of the United Nations and Statute of the International Court of Justice. Retrieved from https://www.un.org/en/about-us/un-charter/full-text (accessed July 2, 2023).

UN COP21 Agreement, 2015. United Nations Treaty Collection. Published December 12, 205. Retrieved from https://treaties.un.org/pages/ViewDetails.aspx?src=TREATY&mtdsg_no=XXVII-7-d&chapter=27&clang=_en (accessed August 26, 2023).

UN IPCC AR4, 2007. AR4 Climate Change 2007: The Physical Science Basis. United Nations Intergovernmental Panel on Climate Change. Retrieved from https://www.ipcc.ch/report/ar4/wg1/ (accessed August 29, 2023).

UN IPCC Mission, 2023. United Nations Intergovernmental Panel on Climate Change. Retrieved from https://www.ipcc.ch (accessed August 29, 2023).

UNEP–Strong, 2023. Maurice F. Strong. UNEP Executive Director, 1972 to 1975. United Nations Environment Programme. Retrieved from https://www.unep.org/unep-50-leaders-through-years/maurice-strong (accessed September 8, 2023).

US CBO Electric Grid, March 2020. Enhancing the Security of the North American Electric Grid. Congress of the United States Congressional Budget

Office document https://www.cbo.gov/system/files/2020-03/56083-CBO-electric-grid.pdf (accessed January 10, 2023).
US EIA AER, 2011. Annual Energy Review, Appendix E, United States Energy Information Administration. Retrieved from https://www.eia.gov/totalenergy/data/annual/archive/038411.pdf (accessed October 9, 2023).
US EIA MER Table 1.2, 2022. August Monthly Energy Review. United States Energy Information Administration. Retrieved from https://www.eia.gov/totalenergy/data/monthly/ (accessed September 27, 2022).
US EIA MER Table 1.3, 2023. September Monthly Energy Review. United States Energy Information Administration. Retrieved from https://www.eia.gov/totalenergy/data/monthly/ (accessed October 9, 2023).
US EPA EV Myths, 2023. Electric Vehicle Myths. United States Environmental Protection Agency. Retrieved from (updated August 28, 2023) https://www.epa.gov/greenvehicles/electric-vehicle-myths (accessed October 14, 2023).
Villalovis, E., 2023. Barack Obama Has Been Working for Months to Shape the White House's AI approach. Washington Examiner. Published November 3, 2023. Retrieved from https://www.washingtonexaminer.com/policy/technology/biden-enlisted-obama-help-iron-out-new-ai-order (accessed November 5, 2023).
WCED, 1987. *Our Common Future*. World Commission on Environment and Development, Chairwoman G. Brundtland. Oxford University Press, Oxford, United Kingdom.
Weber, E., 2021. How EV Charging Stations Work? Know Everything Here. EV Observed. Published February 27, 2021. Retrieved from https://evobserved.com/how-ev-charging-stations-work/ (accessed October 14, 2023).
WEF Commons Project, 2023. The Commons Project. World Economic Forum. Retrieved from https://www.weforum.org/organizations/commons-project (accessed September 24, 2023).
WEF GAEA, 2023. New Initiative to Help Unlock $3 Trillion Needed a Year for Climate and Nature. World Economic Forum. Press release Published January 17, 2023. Retrieved from https://www.weforum.org/press/2023/01/new-initiative-to-help-unlock-3-trillion-needed-a-year-for-climate-and-nature (accessed September 24, 2023).
WEF History, 2023. About History. World Economic Forum. Retrieved from https://www.weforum.org/about/history (accessed September 24, 2023).
WEF Mission, 2023. Our Mission. World Economic Forum. Retrieved from https://www.weforum.org/about/world-economic-forum (accessed September 24, 2023).
WEF RF, 2023. Rockefeller Foundation. World Economic Forum. Retrieved from https://www.weforum.org/organizations/the-rockefeller-foundation (accessed September 24, 2023).

Whiting, K. and J. Wood, 2021. Two years of COVID-19: Key Milestones in the Pandemic. World Economic Forum. Retrieved from https://www.weforum.org/agenda/2021/12/covid19-coronavirus-pandemic-two-years-milestones/ (accessed October 25, 2023).

Whitman, A., 1974. Walter Lippmann, Political Analyst, Dead at 85. Retrieved from https://www.nytimes.com/1974/12/15/archives/walter-lippmann-political-analyst-dead-at-85-walter-lippmann.html (accessed March 13, 2018).

Wigley, T.M.L., R. Richels, and J.A. Edmonds, 1996. Economic and Environmental Choices in the Stabilization of Atmospheric CO_2 Concentrations. *Nature*, 18 January, pp. 240–243.

Woolf, L.S., 1916. International Government. Brentano's, New York. Retrieved from https://ia600901.us.archive.org/23/items/internationalgo00commgoog/internationalgo00commgoog.pdf (accessed July 22, 2023).

Yasmin F., H. Najeeb, U. Naeem *et al.*, 2023. Adverse Events Following COVID-19 mRNA Vaccines: A Systematic Review of Cardiovascular Complication, Thrombosis, And Thrombocytopenia. *Immunity, Inflammation and Disease* 11(3), e807.

Yergin, D., 1992. *The Prize*. Simon & Schuster, New York.

Yergin, D., 2020. *The New Map — Energy, Climate, and the Clash of Nations*. Penguin Press, New York.

Index

A
adaptation, 75–76, 84, 89, 95, 207
Africa, 36, 136–137, 169, 183
Africa, East, 92
Africa, North, 155–156
Africa, South, 136, 138, 168, 171–172, 185
African American, 154
Agenda 21, 96–99
American Marxism, 152, 196
anarchy, 32, 211
anthropogenic climate change (ACC), 71, 75–76, 81–83, 85–90, 100, 103, 105, 113–114, 171, 212
anthropogenic global warming, 88
Arctic, 71
Asia, 5–6, 12, 33, 36, 142, 169, 170
Australia, 33, 199–200

B
Belt and Road, 168–169, 210
Biden Administration, 109, 160, 178, 194, 196
Biden, Joe, 82, 86, 108–109, 160, 190, 193, 195, 197
Biden/Harris, 109
Big Cycle, 205–207, 209

Bilderberg group, 122
Brazil, 86–87, 171–172
BREXIT, 212
BRICS, 168, 171–172, 206, 210
Brooks, Arthur, 57
Browner, Carol, 104
Brundtland, Gro, 96, 100
Bureaucratic Ruling Class, 133, 140–141
Burnham, James, 133, 141–142

C
Canada, 91–94, 96, 138, 160, 171, 200
capitalism, 9, 45–46, 140–142, 154, 202
Carnegie, Andrew, 117, 122
carbon capture and storage (CCS), 74
carbon capture, utilization and storage (CCUS), 74
carbon tax, 87
cell biology, 178–179
central planning, 120, 212
Chaffetz, Jason, 133, 144
Chaos, 32, 212
China, 5–6, 10–12, 21, 23, 32–33, 63, 77–81, 86, 93, 101, 107, 137, 145,

168–172, 182–183, 192, 197–198, 209–210
Christian, 42, 147–149, 151–154
civil rights, 168, 173, 179, 186, 197, 201
civilization, 32–36, 89, 98, 186, 203
Clash of Civilizations, 31, 35
Clash over Resources, 31, 35, 211
climate change debate, 83, 88
climate model, 84, 85, 90
CLINTEL, 89–90
Club of Rome, 123
Cold War, 31–32, 34, 42, 65, 118, 203
combustion, 73, 84, 87, 106–107, 127, 157, 218
communism, 42, 46, 93, 151, 154–155
conservation, 36
conservative, 120, 152, 201
conspiracy, 134, 154–155
COP21, 80–82, 85, 106
Council on Foreign Relations
Covid, 78, 107, 113, 146, 170, 173, 178–179, 181–187, 191–193, 195, 197–202, 210, 215, 217
Covid-19 Pandemic, 78, 107, 113, 173, 178, 181–182, 184, 186, 191, 193, 195, 197, 199, 201, 210
Croly, Herbert, 59–61
Cultural Marxism, 152
currency, 163–164, 172, 201, 206, 208, 210

D

Dalio, Ray, 205–209
Davidson, Thomas, 52, 55
decarbonization, 85
Declaration of Independence, 45, 149
DEI, 168
Deep State, 131, 143–145

democracy, 8–9, 25, 27, 29–31, 123, 131, 194, 196, 203
Democrat, 103, 109, 145, 151–153, 178, 187, 190, 193, 195–197
dialectical materialism, 43–44, 47, 141–142
DNA, 180–181

E

Earth Charter, 99–100
ecology, 94
economics, 29, 59–60, 119–120, 127, 152, 165
economies, 13, 36–37, 77, 81, 107, 127, 171–172, 202–203, 206, 218
efficiency, 98, 105, 114, 167
Egypt, 136, 162–167, 172
Eisenhower, Dwight, 131–133, 143, 193
electric grid, 19–20
electric power, 21
electricity generation, 74, 155–156, 159, 215
emissions, 72–75, 77–83, 86–87, 106, 167
empires, 8, 25–26, 196, 205–207
empires, fate of, 25
empires, lifetimes of, 25–26
energy carrier, 127, 156
energy conservation, 36
energy consumption, 31, 78, 215–218
energy mix, 23, 213, 215, 217
energy production, 107–108
Engels, Friedrich, 43–44, 46–47, 51
environmental impact, 46, 217
environmental socialism, 41, 71, 91
environmentalism, 39, 42, 47, 62, 67, 103
ESG, 167–168, 202–203
Establishment Clause, 149–150
EU 2050, 155, 159–161

Europe, 4, 6–7, 12, 16, 33–34, 46–47, 55, 57, 65, 92–93, 120–123, 131, 139, 142, 152, 155–156, 159–160, 169, 174, 203
European Union (EU), 8, 23, 32, 36, 80, 155, 171, 212
evolution, 25, 44, 95, 115, 134, 181, 206–207
evolutionary, 44, 51, 207

F
Fabian Society, 48, 51–56, 59–60, 120, 142
Fabian Socialism, 48, 51, 58, 143, 155
fascism, 120
feudal, 6, 174, 212
feudalism, 212–213
forecast, 73, 123, 215, 217–218
fossil energy, 87
fossil fuels, 3, 36–37, 74, 86–88, 98, 106, 108, 156, 159, 161, 211–212, 218
France, 4, 7, 12, 21, 23, 25, 32, 58, 61, 63, 86, 148, 171–172, 200

G
GAEA, 128–129
GAIA, 76
Gaza, 163, 165–167
geologic sequestration, 73–74
geopolitics, 36, 209
geopolitical paradigm, 31
Germany, 4, 7–9, 11–13, 16, 32, 36, 47, 58, 61–62, 120, 137, 160–162, 171
glacier, 76–77, 84
glaciation, 76
global government, 41–42, 57, 67, 71, 101, 113
global warming, 77, 86, 88–89, 103
globalism, 8, 54–55, 57, 67, 212

globalization, 63, 113, 204, 210–212
global currency, 172, 210
Glubb, John Bagot, 25–27, 34
Gore, Al, 103–104
Goldilocks Policy, 3, 83, 85, 213, 215, 217–218
Great Britain (or Britain), 7, 19, 26, 40, 47, 53, 58, 63, 86, 137, 156, 161, 164
Great Reset, 201–203
greenhouse effect, 75
greenhouse gas, 72–73, 75, 77, 80–81, 83, 85, 87, 106–107
greening earth, 73
Gross Domestic Product (GDP), 97, 145–146
guano, 46

H
Hamas, 161, 165–167, 218
Hardie, Keir, 52–54
Harris, Kamala, 153
Hawaii, 12, 17, 72
Hayek, Friedrich, 120, 212
Hegel, Georg, 42–43, 45, 63
Holdren, John, 104
Holland, 169
House, Edward M., 58, 60, 139, 140
Huntington, Samuel P., 31–32, 36
hydrate, 74–75
hydropower, 174

I
India, 23, 77, 80–81, 138, 171–172, 183
Inquiry, The, 57, 58, 60, 139, 140
internal combustion engine (ICE), 127, 157, 218
internationalism, 119, 124, 139
International Thermonuclear Experimental Reactor (ITER), 22–23, 219

IPCC, 84–85, 89, 103
Iran, 35, 162, 166, 172, 210
Israel, 161–167, 200, 218

J
Jackson, Lisa, 104
Japan, 5–7, 10–13, 16–18, 23, 33, 76, 80, 122–123, 137, 161, 171, 183
Jewish, 42, 162–163
Johnson, Manning, 154

K
Keeling, Charles, 72
Keeling curve, 72
Kerry, John, 81
Klare, Michael, 36–37
Koonin, Steven, 85
Korea, 6, 23
Korea, North, 101
Kyoto Protocol, 76–77

L
Labour Party, 10, 52–54, 198
Lankester, Edwin Ray, 47–48
Laws of Matter, 44
League of Nations, 8, 11, 41–42, 55–57, 60–62, 133, 139–140, 162, 208
Lenin, Vladimir, 10, 48–49, 51, 62–63
liberal, 53, 59, 94
Liebig, Justus von, 46–47
Lippmann, Walter, 58–62
lockdown, 170, 183–184, 191–192, 194, 197–199, 201–202
Locke, John, 45
Lomborg, Bjorn, 86–87
London School of Economics (LSE), 59, 120

M
MacDonald, Ramsay, 53–54
Malaysia, 33

Managerial Revolution, 133, 141–142
managerialism, 142
mass media, 134
materialism, 27, 43–45, 47, 125, 134
Mayflower Compact, 147–148, 150
Marx, Karl, 42–43, 47, 51–52, 62–63
Marxism, 120, 141, 152, 196
McCarthy, Regina, 104–105
Mexico, 33, 86
Middle East, 5, 8, 25, 34, 161–163, 169
military–industrial complex, 131–133, 193
Milner, Alfred, 47, 138
monarchy, 6, 9
monetary policy, 121, 208
mRNA, 179, 181, 185, 187, 192–193
multicultural, 33–34

N
natural resources, 10, 36, 46, 48–49, 67, 96–97, 105, 123, 171, 211
New Republic, 59–61
New Zealand, 33, 198, 200
NIPCC, 84–85, 89
North America, 6, 33, 122–123, 203
Nuclear Age, 13–14
nuclear energy, 20–21
nuclear fission, 3, 12, 16, 18, 20–21, 87, 108, 156, 159, 161, 215, 217
nuclear fusion, 18, 20–23, 87, 218–219

O
Obama Administration, 108–109, 144, 189
Obama, Barack, 71, 103–104, 108–109, 178, 187
Obama-Biden, 81–82, 104, 106, 108, 160, 187

oil crisis, 161–162, 165
oil price, 161, 163, 165
Oil and Gas Climate Initiative (OGCI), 85–86
oligarchy, 133
OPEC, 33, 35, 162–164
Orwell, George, 142–143

P
Palestine, 161–162, 165–166
pandemic, 78, 107, 113, 173, 178–179, 181–182, 184, 186–187, 191–199, 201–202, 204, 210, 215, 217
paradigm, 31–32, 210
Paris Climate Agreement, 81–82, 85, 100
Paris Peace Conference, 61, 139–140
Pease, Edward, 52–55
philanthropy, 116–117, 121, 124, 126
Plato, 43, 135
plutocracy, 133–134
pollution, 66, 72, 81, 105–106, 123, 167
population growth, 35, 99
Prentis Cycle, 25, 27–28
private sector, 103, 125, 129, 133, 144, 174, 193, 202
privileged minority, 95, 113–114, 119
progressive, 44, 51, 59–60, 118, 151–152, 154, 204
property, 18, 43, 45–47, 57, 135, 140, 142, 192
public policy, 132, 217
public sector, 129, 133, 145, 173–176, 178

Q
quality of life, 35, 37
Quigley, Carroll, 133–135, 137–138

R
radiation, 14, 87
Radical left, 151–152, 203
Reagan Administration, 152
Reagan, Ronald, 18, 29, 153
Republic, American, 45, 204
Republic, New, 59–61
Republic, Plato's, 135
Republic, secular, 149
Republic, Weimar, 8–9
Republican, 95, 103, 108–109, 124, 145, 178, 187–188, 193, 195–197
reserve currency, 206
revolutionary, 4, 7, 10, 48, 51, 95, 124
Rhodes, Cecil, 135–138
Rhodes Secret Society, 137–138
Rio Conference, 96–97, 99
Rizzi, Bruno, 133, 140–141
RNA, 179–181
rock oil,
Rockefeller, David, 92, 116, 118–126
Rockefeller, John D., 115–116, 118
Rockefeller Brothers Fund, 121
Rockefeller Foundation, 116–119, 124, 126, 128–129
Roe v. Wade, 152–153
Round Table, 56, 138, 140
rule of law, 101
ruling class, 9, 44, 133, 135, 140, 141, 145, 210, 212
Ruskin, John, 134–135
Russia, 4–10, 19, 21, 23, 26, 32–33, 48–49, 61, 158–161, 169–172, 189, 191, 210
Russia–Ukraine, 155, 158, 160, 171
Russian Federation, 63, 80

S
Saudi Arabia, 35, 86, 164, 166, 172
sequestration, 73–75, 87, 107

serfdom, 120, 212
Sherman Antitrust Act, 115
Skousen, Cleon, 134
Smil, Vaclav, 123
socialist environmentalism, 42
solar, 20, 81, 87, 109, 156, 215
South America, 33, 137
Spain, 26, 86, 156
Stalin, Joseph, 10, 62–63, 140, 151
Stockholm Conference, 67, 94–97, 99
Strong, Maurice, 67, 91–93, 96, 99–101, 103, 113, 118, 199, 122–123
super grid, 155–157
Supranational Authority, 54–55, 57, 61, 210–211, 213
Sussman, Brian, 42, 44, 46
sustainable development, 97–98
sustainable energy, 3, 74, 85, 155, 157–158, 161, 211, 213, 215, 217

T
Tansley, Arthur, 48
temperature, 71–72, 80, 84–85, 217
Thant, U, 41, 65–67, 91, 94, 211
Tragedy and Hope, 133–134
Trilateral Commission, 119, 122–123
Trotsky, Leon, 141–142
Trump Administration, 81–82, 105–106, 108–109, 144, 166, 183–184, 189, 191, 194
Trump, Donald, 81–82, 106, 108, 109, 153, 178, 187–191, 193, 195, 197, 204, 212
Turkey, 7, 33, 61
Tyranny of the Majority, 29–31

U
Ukraine, 10, 19–20, 36, 155–156, 158–161, 171, 190, 218
uniparty, 145
unionization, 173
unions, public sector, 173–174, 176, 178
United Kingdom, *see also* Great Britain
United Nations Environment Programme (UNEP), 84, 96–97, 99

V
vaccine, 179, 184–187, 193
Vita Nuova, 52, 55

W
Wallas, Graham, 58–60
Western Civilization, 32–36, 186, 203
wind, 75, 109, 156, 215
Wilson, Woodrow, 7–8, 55, 60, 62, 118
Wilson's 14-Point Plan, 57, 60–62
Woolf, Leonard, 54–56
World Economic Forum, 119, 126, 128, 193
World War I, 6–10, 13, 25, 41, 48, 54–55, 57–58, 60, 62, 139–140, 161
World War II, 8, 11–13, 15, 27, 31–32, 41, 55–56, 62–63, 65, 118, 121–122, 131, 143, 145, 155, 162, 172

Y
Yergin, Daniel, 37, 114–115, 211
YMCA, 92–93